Tales of Tropical Plant Diseases in an Age of Climate Change

Khoon Chin

Tales of Tropical Plant Diseases in an Age of Climate Change

A View of Sustainability Based on Complexity Science

Khoon Chin
Therwil, Switzerland

ISBN 978-3-031-90789-0   ISBN 978-3-031-90790-6   (eBook)
https://doi.org/10.1007/978-3-031-90790-6

© The Editor(s) (if applicable) and The Author(s), under exclusive license to Springer Nature Switzerland AG 2025

This work is subject to copyright. All rights are solely and exclusively licensed by the Publisher, whether the whole or part of the material is concerned, specifically the rights of translation, reprinting, reuse of illustrations, recitation, broadcasting, reproduction on microfilms or in any other physical way, and transmission or information storage and retrieval, electronic adaptation, computer software, or by similar or dissimilar methodology now known or hereafter developed.
The use of general descriptive names, registered names, trademarks, service marks, etc. in this publication does not imply, even in the absence of a specific statement, that such names are exempt from the relevant protective laws and regulations and therefore free for general use.
The publisher, the authors and the editors are safe to assume that the advice and information in this book are believed to be true and accurate at the date of publication. Neither the publisher nor the authors or the editors give a warranty, expressed or implied, with respect to the material contained herein or for any errors or omissions that may have been made. The publisher remains neutral with regard to jurisdictional claims in published maps and institutional affiliations.

This Springer imprint is published by the registered company Springer Nature Switzerland AG
The registered company address is: Gewerbestrasse 11, 6330 Cham, Switzerland

If disposing of this product, please recycle the paper.

For Twannie

Foreword

Plant pathologists like Dr. Khoon Chin, who study the causes and control of diseases that afflict tropical crop plants, rank high among the unsung heroes of modern times.

The range of crops they must work with is immense, including, among many others, grains such as rice, maize, and sorghum, field crops like soybeans and cassava, and plantation crops that include bananas, oil palm, and rubber. Most are staple food sources or vital tradeable commodities that help to sustain the food security and economic stability of the countries where they are grown and with whom they are traded. Moreover, the range of organisms that cause disease on such crops is even more diverse, including many species and strains of fungi and fungus-like organisms, bacteria, and viruses, many also involving carriers (vectors) such as insects or even humans. When combined with the range of potential control strategies that include breeding or engineering genetically resistant cultivars, the deployment of safe chemicals, or the use of mixed cropping systems, the complexity mounts. It doesn't stop there, however, for global environmental changes, especially climate change, may alter the dynamics of some diseases, or a breach of biosecurity may result in the import of an entirely new, aggressive pathogen. Either scenario, as also a sudden genetic change in the aggressiveness of an existing pathogen, could lead to an epidemic resulting in widespread starvation or the collapse of a once stable economy, or both. Finally, further levels of complexity result from the need to increase the diversity of crops grown, the requirement that control measures used do not damage the biodiversity of natural ecosystems, and the ever-present yet unpredictable prospect of warfare or a natural disaster.

During a long and highly successful career in the study and management of tropical plant diseases, Khoon has learned how to take such complexity in his stride and in *Tales of Tropical Plant Diseases in an Age of Climate Change* uses his experience-based wisdom to enable others—students and a wider public alike—to follow in his footsteps with confidence. The book, as its title suggests, is written in a very personal, exceptionally lucid narrative style that Khoon uses with great skill to draw his reader into his complex world and as they go on to gradually enlarge it and make it their own. This is not a private world, however, for by constantly and perceptively taking a global view of his subject, Khoon helps his readers to understand how each tropical crop and its diseases fits into an ever-changing worldwide picture of the historical, current, and potential food and bio-commodity production and consumption.

This outstanding book is at once both immensely enjoyable and highly informative; it deserves a very wide readership.

David Ingram, OBH, VMH, PhD, ScD, FLS, FRSB, FRSE
St Catharine's College, University of Cambridge, Cambridge, UK (formerly Master)
Royal Botanic Garden, Edinburgh, UK (formerly Regius Keeper & Director)
University of Lancaster, Lancaster, UK (Honorary Professor)
University of Edinburgh, Edinburgh, UK (Honorary Professor)
8 January 2025

Preface

Although the importance of agricultural crops in providing sustenance to the world's population is obvious, threats to their productivity from disease, pest, and environmental extremes are often far less well known, and for most people a risk we gladly relegate to our farmers to manage. The threat to food security from damage to crops grown in the tropics is, in particular, often underestimated. As a result, there could be a political risk in the future of under-investment in technologies that are needed to keep these threats at bay.

Interest in the cultivation of tropical crops is also likely to increase, as climate change enables their cultivation at greater distances from the equator, or limits cultivation in existing areas.

This science book brings together in a single volume, a global perspective of major tropical crops, their role in food production, and the threats to their productivity that are posed by plant diseases. It is intended to reflect on the author's personal experiences and perspectives of professional plant pathology rather than just be an encyclopedic collection of facts. Unusual for a scientific text, it has been written largely in the form of narratives and pictures to reflect current trends in collaborative science and self-learning in education. Standards of scientific rigor and referencing have, however, been maintained.

Narratives have always been humans' favored way of communication (Snowden 2003). Whether they are a written couple of words or pages long, spoken or drawn, narratives are arguably the most knowledge dense of our communications with each other.

There are therefore three simple reasons why I have written the book in this manner. Firstly, by reflecting on the stories of others the professional scientist may, according to complexity theory, be stimulated to form new perspectives or unique insights based on his/her *own* experiences and knowledge (see also Chap. 2). Some knowledge management circles describe the process as helping people "know what they did not know they knew." Complexity practice also suggests that it could help people "know what they did not know they did not know," which are crucial insights for enabling scientific research as opposed to scientific investigation (see Chap. 1).

Secondly, students today tell us that they learn more from the experiences of their teachers (and their own) than from just facts. Most classic texts are encyclopedic and didactic in nature leaving little room for modern practices of self-learning. Stories are on the other hand, arguably the most effective way of communicating experiential/tacit knowledge. We are currently at the cusp of rapid progress in Generative AI, which suggests that the role of textbooks may need to change (see Chap. 8).

Thirdly, it is written to be "light in text and heavy in knowledge." The use of narratives and pictures provide the bonus of improving accessibility of the contents to the non-professional reader who has interests in current topics like sustainability, biodiversity, food and biosecurity, regenerative agriculture, resilience, ecosystem services, climate change, investments in new technology and public policy. For rapid access to specialist information, the narratives are balanced by an extensive appendix containing a library of referenced facts about the management of tropical crop diseases.

Framed by the two major trends of our time, Climate Change and Artificial Intelligence, the concluding chapters bring together elements of the journey to bear on three keynotes for the future, sustainability, adaptability, and resilience. New approaches are proposed for studying climate change based on complexity theory. Finally, they summarize our journey into a single narrative with a reminder of how this book could be used (Chap. 9, Epilogue).

Texts that are comparable in scope have previously been written. The book is intended to update older texts where appropriate and to complement current ones where they are more limited in scope.

To fellow plant pathologists and crop protectionists, I hope that you will find this book enjoyable and an interesting read, to be like a visit to old friends, and perhaps discover a way to make some new ones. To other readers, I also hope that these narratives may have provided you with some new perspectives on food production in the tropics that are relevant to your interests, as well as an introduction to complexity theory for future studies.

Therwil, Switzerland
1 March 2025

Khoon Chin, PhD

Reference

Snowden DJ (2003) Narrative patterns: the perils and possibilities of using story in organisations. In: Lesser E, Prusak L (eds) Creating value with knowledge. Oxford University Press, 14p

Acknowledgments

I am indebted to Prof. Arthur H. Bunting (formerly University of Reading, UK) for introducing me to the botany of agricultural plants in the tropics, to Prof. David J. Snowden (Center for Applied Complexity, University of Wales, UK) for teaching me most of what I know about the art and science of narratives for managing knowledge, and to Prof. Martin S. Wolfe (formerly ETH Zurich) for teaching me about the role of biodiversity in crops and agroforestry. I am grateful to Prof. David S. Ingram, OBE, FRSE, FCI Hort., FRSB (former Master of St Catherine's College, Cambridge, and Regius Keeper of Royal Botanic Gardens Edinburgh) and Prof. Matt Dickinson (Nottingham University, UK; presently Vice President, British Society Plant Pathology (BSPP) for reviewing this book and encouraging me in this experimental approach to communicating science and facilitating self-learning. I thank Steve West (Head of Global Development, Syngenta, Basel) for helping me clarify key messaging so as to better meet reader needs, Prof. Urs Hilber (Dean, University of Applied Sciences, Zurich) and Dr. John F Scheuring (formerly Plant Breeder International Crops Research Institute for the Semi-Arid Tropics (ICRISAT)) for critically reading this document. All errors in interpretation and elaboration are of course entirely my own.

I am grateful to K.S. Teh, H. Baharol, S.A. Bakar, and T.S. Han for warm and valuable technical support during my early "tropical years."

I thank Zuzana de Ruiter, Springer Nature, for understanding the purpose of this book, and for kindly guiding me through the publication process.

Competing Interests The author has no competing interests to declare that are relevant to the content of this manuscript.

Contents

1 **Introduction**... 1
   1.1 Why I Wrote This Book............................ 2
       1.1.1 The Threat 2
       1.1.2 Storypad 1: A Brief History of the Impact of Some Diseases of Crop Plants on the Affairs of Man 2
       1.1.3 The Gap in Societal Appreciation of the Threat 5
   1.2 Why Bother Specifically About Tropical Crops?............... 6
   1.3 Why a Narrative Approach Embedded in Complexity Theory? ... 8
   1.4 How Is the Book Organized?............................... 9
   References.. 10

2 **Time to Check Our Bearings Before We Start On Our Journey** 13
   2.1 Crop and Disease Trends Over the Period 1957 to 2022......... 13
   2.2 Tales from Some of the Crop and Disease Trends.............. 14
       2.2.1 Crop Trends (see Fig. 2.1) 15
       2.2.2 Disease Trends (Fig. 2.2) 16
       2.2.3 Tale of the Ubiquitous 3 in the Tropics 17
       2.2.4 A Gentle Reminder 17
   Reference 17

3 **Cereals** 19
   3.1 Rice 19
       3.1.1 Production 19
       3.1.2 Storypad 2: 'Tall Tales' of Rice: The Astonishing Diversity................................... 20
       3.1.3 Itinerary 20
       3.1.4 Rice Disease Panorama 21
       3.1.5 Storypad 3: Sheath Blight Disease (the 'Painter Who Would Be King')......................... 24
       3.1.6 Tales of Mistaken Identity 26
       3.1.7 Rice Disease Control 27
       3.1.8 Rice Disease Pictures.......................... 28

	3.2	Maize (Corn)	29
		3.2.1 Production	29
		3.2.2 Maize Disease Panorama	29
		3.2.3 Storypad 4. Not All Fungal Pathogens Are Bad News	31
		3.2.4 Maize Disease Pictures	32
	References		32
4	**Field Crops**		35
	4.1	Tobacco	35
		4.1.1 Production	35
		4.1.2 Itinerary	36
		4.1.3 Storypad 5: The 'Black Gold' of the San Andrés Valley, Mexico	36
		4.1.4 Tobacco Disease Panorama	36
	4.2	Groundnuts	37
		4.2.1 Production	37
		4.2.2 A Tale from the Sub-Sahara by Scheuring, JF (pers comm. 2022)	38
		4.2.3 Storypad 6: Ancient Survival Packages Still Relevant Today	38
	4.3	Soybeans	39
		4.3.1 Production	39
	4.4	Mung Beans	40
		4.4.1 Production	40
		4.4.2 Storypad 7: Not Just for Beansprouts	40
	4.5	Cassava	40
		4.5.1 Production	40
		4.5.2 Legume and Root Crop Disease Panorama	40
		4.5.3 Groundnuts diseases	41
		4.5.4 Soybeans diseases	41
		4.5.5 Mung Bean diseases	42
		4.5.6 Cassava diseases	42
		4.5.7 Field Crop Disease Pictures	43
	References		45
5	**Plantation Crops**		47
	5.1	Bananas	47
		5.1.1 Production	47
		5.1.2 Fancy Bananas	48
		5.1.3 Itinerary: Philippines (Fig. 5.1a)	48
		5.1.4 Storypad 8: Two Tales of How a Fungus Determined the Banana We Eat	49
		5.1.5 Banana Disease Panorama	50
		5.1.6 Storypad 9. Aleyandro's Tale: A Day in the Life of a Banana Plantation Manager in Costa Rica	51
		5.1.7 Storypad 10. Size Helps. A Whole Replicate of a Field Trial Can Fit on a Single Leaf	51

	5.2	Oil Palm..	52
		5.2.1 Production...	52
		5.2.2 Oil Palm, Rubber and Cocoa Itinerary................	53
		5.2.3 Oil Palm Disease Panorama..............................	53
		5.2.4 Storypad 11. Pathogen and Panacea...................	53
		5.2.5 Storypad 12. Shining a Light with LAMP: Model Specifications for Field Diagnostic Technology.........	54
	5.3	Rubber...	54
		5.3.1 Production (Table 5.2).....................................	55
		5.3.2 Rubber Tree Disease Panorama........................	55
	5.4	Cocoa..	56
		5.4.1 Production (Table 5.3).....................................	56
		5.4.2 Cocoa Disease Panorama.................................	56
		5.4.3 Storypad 13. The Cryptic Pathogen of Cocoa......	57
		5.4.4 Plantation Crop Disease Pictures (Figs. 5.1, 5.2, 5.3, 5.4, 5.5, 5.6, and 5.7).............	57
	References...		61
6	**Vegetables**...		65
	6.1	Solanaceous Vegetables and Beans.............................	65
		6.1.1 Production (see Table 6.2)...............................	66
		6.1.2 Solanaceous Vegetables and Beans Disease Panorama....	66
		6.1.3 Storypad 14: A Tiny Pest of Superlative Proportions.....	67
	6.2	Cucurbit Vegetables..	68
		6.2.1 Cucurbit Disease Panorama.............................	68
		6.2.2 Other Vegetable Diseases Panorama.................	69
		6.2.3 Vegetable Disease Pictures..............................	69
	References...		71
7	**Fruit Crops**..		73
	7.1	Mango...	73
		7.1.1 Production (Table 7.1).....................................	74
		7.1.2 Mango Disease Panorama................................	74
	7.2	Durian...	74
		7.2.1 Storypad 15. The Legendary Fruit of SE Asia and Its Nemesis?................................	75
		7.2.2 Durian Disease Panorama................................	76
	7.3	Other Fruits...	76
		7.3.1 Pineapple: A Fruit Fit for Kings.......................	76
		7.3.2 Papaya (Carica Papaya)..................................	77
		7.3.3 A Tale About Three Apples That Really, Are Not Apples...	77
		7.3.4 Tale of a Hairy Fruit and Its Relatives...............	78
		7.3.5 Fruit Disease Pictures (Figs. 7.1 and 7.2)..........	78
	References...		79

8	**The Future: Sustainability, Food Security and Resilience in an Age of Climate Change**		81
	8.1	A Tale of Freight Train Trends	81
	8.2	Studying the Impact of Climate Change	82
		8.2.1 The Complexity Theory Approach	83
		8.2.2 Geographical Implications	84
	8.3	Sustainability	84
		8.3.1 Storypad 16: Mario's Tale	85
	8.4	Resilience	87
	8.5	IPM	88
		8.5.1 Storypad 17: Food for Thought from 'Systems Studies'	88
	8.6	Technological Advances	90
	8.7	Artificial Intelligence: A Second Freight Train Trend?	91
	References		92
9	**Epilogue: Weaving the Narratives Together**		95
Appendices			97
Disclaimer			119
Index			121

About the Author

Khoon Chin The author holds a BSc Hons degree in Agricultural Botany (Reading University, UK) and a PhD in Plant Pathology and Population Genetics (Cambridge, UK), as well as an MSc of USM, Malaysia. He has had a lifelong global career as a research and teaching phytopathologist, working for national (MARDI, Malaysia; PBI, UK) and international organizations and for industry, in both the tropics and temperate regions. He was formerly a Bye-Fellow of Downing College, UK, Visiting Scientist at the International Rice Research Institute, the Philippines, lecturer at the Bern Fachhochschule for Agriculture (HAFL, Switzerland), and an editor of *Plant Pathology* (Blackwells) Journal of the British Society of Plant Pathology. He has written well over a hundred publications in refereed journals and lectured widely as an invited/keynote speaker in international and national symposia in EU, the USA, Asia, and Latin America. He is also accredited to the Cynefin Center for research on complexity science and was for 8 years an independent consultant for complexity areas of business management.

Chapter 1
Introduction

Abstract The global impact of the recent COVID-19 pandemic has been a sharp reminder of the fragility of human life, which is under constant threat from an ever-changing pattern of disease evolution. Perhaps it is also a reminder that the food and feed crops on which we and our domestic animals ultimately depend are equally at threat from pests, diseases and environmental extremes.

It would be no exaggeration to argue that our survival as a species in the future will partly depend on our ability to protect these and related crops from maladies that affect their productivity. Taken against a backdrop of growing nutritional requirements, changing expectations and the dwindling availability of arable land and water made worse by climate change, the threat of disease adds complexity to the key challenges in food security: sustainability, access and utilization. In medicine, we have only to manage diseases of a single host species, humans, but plant pathology has the task of shielding multiple species of crop plants, requiring the management of multiple pathosystems (pathogens, their vectors and hosts). Another critical difference is the inability of crop plants to 'complain' to us when something is amiss, until it's too late and damage has already been sustained.

Theory (see Chap. 8) tells us that complex problems cannot be easily solved by 'best practices' based on historical experiences because the complex domain is not rule-based (does not follow linear causality) and has too many variables to effectively model. They require a different framework for study, to make sense of and design measures to exploit or mitigate. We need new ways of helping us get beyond the mindset of learning about 'what we know we do not know' to 'what we do not know we do not know'. That is the complexity challenge that drives this book.

1.1 Why I Wrote This Book

1.1.1 The Threat

Most people are less familiar with the idea that crops suffer from disease just as much as humans and our domestic animals do. Past epidemics, however, attest to the ability of plant pathogens to greatly influence the affairs of man, from famine and human migrations to economics and cultural/dietary habits. Storypad 1 lists some of the stories embedded in our plant pathology communities that exemplify some of the challenges we have faced in the past. The narratives do not necessarily represent diseases with the greatest economic impact but are selected to represent the range of threats.

They do, however, speak to us about the resilience of several of these pathogens, which have co-evolved with our crop plants. Despite the best efforts of man to control it, blast of rice has, for example, remained till today perhaps the most serious disease on the crop since records were first made in the early seventeenth century in China (see rice in storypad 1). The threat caused by stem rust of wheat has been known since the time of Aristotle but is set for a comeback because of the return of favourable conditions (climatic change, evolution of new races and increasing incidence of the alternate host barberry; Schumann and Leonard 2001).

1.1.2 Storypad 1: A Brief History of the Impact of Some Diseases of Crop Plants on the Affairs of Man

Wheat
Santini et al. (2018) quoted reports by Roman authors Cicero, Varno and Columnella (2100–1950 BP) of stem rusts of wheat (caused by Puccinia graminis), so feared that a god/goddess of rust (Robigus/Robigine) had to be appeased to prevent damage.

Potato
The Irish Potato Famine in the 1840s was caused by crop failure over successive years due to an epidemic of late blight disease (caused by Phytophthora infestans) of potatoes in Ireland and some other parts of Europe. As a result, a million people died of starvation in Ireland and another million migrated to the United States ('the potato plants rotted as the Irish peasants watched helplessly'; Schumann and D'Arcy 2017).

Rice
About 2 million people died of starvation during the great Bengal famine in 1943. It was triggered by a crop failure caused by brown spot disease (causal agent Bipolaris oryzae, teleomorph Cochliobolues miyabeanus) of rice (Padmanabhan 1973).

The most important disease of rice today, however, is blast (causal agent Magnaporthe grisea). Known since the late Ming Dynasty in China as 'rice fever disease' (Ou 1985), it causes damage throughout most of the rice-growing world. Based on FAOSTAT estimates of crop losses in 2017 and a conservative estimate that 10% of this damage to rice was due to blast, it can be estimated that the damage caused could have fed 227 million more people in that year.

Coffee
The coffee tree is native to Africa, but coffee rust (caused by Hemileia vastatrix) has restricted its cultivation there, in Sri Lanka, and in other countries in SE Asia since the 1890s (Talhinhas et al. 2017). The disease has since spread to the new world, but the delay has helped Brazil and other Latin American countries today to become more successful than Africa and other competitors in producing coffee. Another contributory reason has perhaps been the cultivation of coffee at high altitudes with a distinct dry season in Latin America, which has favoured the successful use of fungicides against the disease.

Tea
Why the British drink more tea than coffee may be connected in part, to the decimation of Sri Lanka's coffee plantations by rust disease (see above) when it was still a colony of the United Kingdom (Webb 2002). The disease led to the rapid replacement of coffee plantations with tea in that country.

Rubber
The rubber tree is native to South America, but South American leaf blight (caused by *Pseudocercospora ulei*) has prevented all attempts to establish commercial plantations in that subcontinent from 1933 to 2002 (Liberei 2007). As a result, Malaysia and, more recently, Indonesia have taken the global lead in the production of natural rubber by implementing strict phytosanitary measures against the disease.

Cocoa
Cocoa is native to the rainforests of Central America. In a natural forest, the diversity of plant species prevents the rapid spread of diseases and pests between neighbouring plants. Under the monoculture of plantation conditions, diseases like Witches broom caused by Moniliophthora perniciosa (previously Crinipellis perniciosa), and frosty pod rot (caused by the related Moniliophthora rorei) have become serious problems in Latin America since the end of the nineteenth century (de Souza et al. 2018). This threat provided a competitive advantage for countries in Africa and Asia that are free of the disease, allowing them to become the current global leaders in the amount and quality of cocoa produced.

Banana
The Panama disease of bananas in Latin America (caused by Fusarium oxysporum) is why the Western world at present eats almost exclusively Cavendish cultivars of banana, rather than Gros Michel. The latter was previously the industry favourite because of its fruit characteristics, till the 1950s, when the disease destroyed large areas of the crop and had to be replaced by Cavendish (Ploetz 2005). *On the other hand, Cavendish has not done as well in SE Asia because the presence of the TR4*

race of the Panama disease there hampered its cultivation (see Banana and Fruits trend for an update).

Maize

The effects of biodiversity loss in a crop can be catastrophic because it increases vulnerability to diseases that can thrive on the narrow genetic base. A southern corn leaf blight epidemic (caused by Cochliobolus heterostrophus) destroyed 15% of the yield in the corn belt of the United States in 1970, mainly due to 90% of the maize plants carrying in common the Texas male sterile cytoplasm, cms-T (Bruns 2017). *The damage is estimated to be equivalent to ≥ $6.0 billion by 2015 standards. Removing this single genetic vulnerability by stopping the use of cms-T produced dramatic results in restoring the health of the crop in the United States.*

Soybean

Soybean production in Brazil increased almost nine-fold from 1991 to 2021 (Colussi and Schnitkey 2021a, b), *surpassing the United States in 2018. This spectacular increase in production would have enabled it to overtake the United States earlier, but for the occurrence of Asian Soybean rust, Phakopsora pachyrhizi (first reported in Brazil in 2001), which can reduce yields by 90% and is able to survive throughout the year in Latin America. New challenges like stem canker and seed rot are already appearing on the horizon* (Mena et al. 2023; Dos Santos et al. 2024).

Olive Trees

The sudden wilt of olive trees caused by Xylella fastidiosa subsp. pauca, which has destroyed millions of olive trees in Apulia (reported since 2013 in Italy) and also threatens other European and Mediterranean countries, is an example of how a pathogen not only affects a crop but also the landscape of a region (Schneider et al. 2020).

Citrus

The citrus yellow dragon disease (Huanglongbing or HLB, aka as citrus greening) is caused by the bacterium 'Candidatus Liberibacter asiaticus' and is transmitted in warm climates by the insect vector Asian citrus psyllid. It is probably the most destructive disease of citrus plants. No effective curative measures are available against HLB. For citrus production areas without HLB or with low HLB disease incidence, removal of 'C. Liberbacter asiaticus' inoculum is critical to prevent further HLB spread. Chemical control of the vector is complicated by the long latency period of the disease. HLB is now global in distribution and recently posed a major threat to the citrus industry in Florida (Alvarez et al. 2016), *causing > 75% loss in gross citrus production, as well as in California* (Milne et al. 2018), *where strict quarantine measures are in place.*

Coconut

Lethal yellowing is perhaps the most serious disease of coconuts. It is caused by Phytoplasma palmae, a bacterium spread by sap-feeding insects. Millions of palms have been destroyed in the Caribbean, South America and Africa in the last 20–30 years (Gurr et al. 2016). *Coconuts are crucial to the economics and culture* of local communities in the tropics, so the impact of lethal yellowing can be considerable. Because the pathogen is not easily cultured and is symptomless during incubation, biosecurity measures based on accurate, real-time DNA/PCR methods of detection are critical to prevent its spread to the Oceania area, where it is still largely absent, and its control in areas (e.g. Ghana, Mozambique and Tanzania) where it has already caused severe damage (Mpunami et al. 1999; Dickinson 2015).

According to fossil records, plant parasites, especially obligate parasites that are only able to survive on their living hosts, probably co-evolved with their hosts millions of years ago. Each wild host relative is thought to harbour several hundred species of endophytic and plant pathogenic fungi, so it is reasonable to assume that the threat to plant health is diverse and that many pathogen species may indeed have been distributed by human activity in selecting and transporting plants with desired traits around the world (Santini et al. 2018).

Another potential threat is the introduction of crops to new geographical regions. Crops that are exposed for the first time to new pathogens (e.g. beans and tobacco encountering new viruses harboured by weeds) could lack the defenses that native crops had evolved to survive (Gilbertson et al. 2021). It is therefore important to trace the genetic origin of crops and how they have spread around the world to understand the threats they face.

The challenge posed by crop disease is made more difficult because, for perceived reasons of production efficiency, crops tend to be grown as a monoculture in genetically homogeneous, large, contiguous areas. Such conditions favour the unimpeded spread of epidemics once pathogens take hold. Also, unlike animals and humans, which have an adaptive, circulating system for immune response, plants do not. The opportunities for immunizing plants against attack are currently limited, e.g. to SAR (Systemic Acquired Resistance, Kessmann et al. 1994) responses.

The pathology of crop plants does, however, have the advantage over human pathology in that crops can be bred for resistance to pathogens.

1.1.3 The Gap in Societal Appreciation of the Threat

This gap in societal appreciation of the importance of crop diseases and pests is unlikely to narrow soon, as new generations of consumers increasingly tend to associate food more with buying in supermarkets than with growing on farms. Solutions for crop disease management could become a lower political priority, with the potential downside that insufficient resources will be allocated for their study and application in the future.

As a former research and teaching phytopathologist, my previous jobs have happily provided me with the ways and means to travel the world, visiting growers, learning and assisting where I could in mitigating their problems. The objective of this book is to share my personal experiences of crop, pathogen and environment (phytopathologists call this 'the disease triangle') in a visual, perhaps slightly touristic way, so as to better gain the attention of non-specialists. The crops and diseases included are therefore not meant to be in any way exhaustive.

Ironically, the symptoms that diseases etch into host plants, and the organisms themselves, are sometimes interesting images of nature in microcosm, which can also make them memorable to study. Two examples are sheath blight of rice (see 'the painter who would be king') and VSD (see 'the cryptic disease of cocoa').

Pathogens and their hosts have co-evolved over time, so their histories are intertwined. Human activity in the domestication and global distribution of our key food crops also inadvertently spread the accompanying pathogens (Santini et al. 2018), adding another layer of complexity to their evolution.

Much of my career has been focused on formal, technical publications, including almost a decade as an editor of an international scientific journal, so this book also represents an attempt to make my work more readily accessible to non-specialists.

Finally, it is hoped that this book could perhaps serve as an abbreviated resource for people with interests in climate change, food security, sustainability, biodiversity, invasive species and biosecurity (also see summary in Chap. 8 on the future).

1.2 Why Bother Specifically About Tropical Crops?

The FAO estimated that in 2020, about 780 to 811 million people in the world faced hunger, with the grim forecast that the number was expected to continue to increase in the coming years. Crops produced in the tropics will have a decisive role to play if we are to mitigate against this trend.

Out of over 50,000 plant species that the world uses for food, just three species, rice, maize and wheat, provide 60% of the world's food energy needs, and 15 species provide 90% of its needs (National Geographical Society 2011; FAO et al. 2021). Ten of these fifteen crops, including rice and maize, were domesticated (Fig. 1.1) and are produced mainly today in the tropics and subtropics (Table 1.1). Rice alone is not only the staple food of about half of the world's population but also that of the poorer half.

For the purposes of clarity, I shall in this book define tropical crops as those which are largely grown in the tropics, as there are temperate crops that can be grown in the tropical highlands, as well as cultivars of tropical crops that can be grown in warm temperate or protected environments in temperate regions.

In this regard, interest in the cultivation of tropical crops is likely to increase as climate change enables their cultivation in more northerly regions.

Diseases and pests are potentially a greater threat in tropical rather than in temperate regions. The annual winter break in the cropping cycle of temperate areas obviously imposes limits on pest and disease growth. So long as water is not limiting, there is not a comparable restriction in the tropics. Crop rotation is only

1.2 Why Bother Specifically About Tropical Crops?

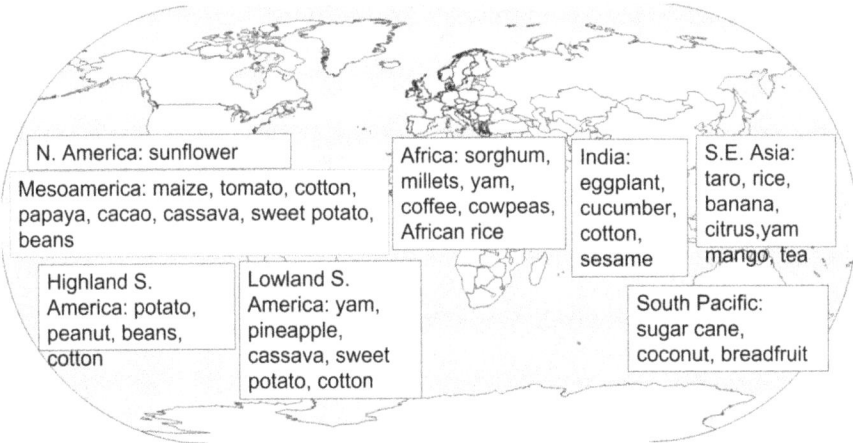

Fig. 1.1 Domestication of major tropical crops. (Source: Harlan 1976)

Table 1.1 The fifteen major food crops of the world and where they are mainly grown

Crops	Tropics	Temperate regions
Cereals	Rice	Wheat
	Corn	Barley
	Sorghum	
Sugar	Sugar Cane	Sugar Beet
Root	Sweet Potato	Potato
	Cassava	
Legumes	Soybean	Beans
	Peanuts	
Tree	Coconut	
	Banana	

possible in space—crops must be separated by sufficiently large distances. In temperate areas, crop rotation is possible both in space and over seasons, thereby allowing soil conditions to recover and specialized pathogen populations to decline during the cold months.

Agriculture in much of the tropics tends to be characterized by small farm sizes and non-uniform planting cycles. These conditions encourage the survival of pests and pathogens between crops and make the timing of farm operations to coincide with low pest levels more challenging.

High temperatures and humidity in much of the tropics also tend to lead to rapid decomposition and loss of organic matter from the soil, with negative effects on soil health and stability.

1.3 Why a Narrative Approach Embedded in Complexity Theory?

My experience with corporate Knowledge Management towards the end of my career suggested that knowledge is best communicated and transferred by relating short, real-life anecdotes, preferably illustrated.

Knowledge sharing and communication within human communities have always been driven by storytelling. Today, narrative has been harnessed by organizations and contemporary media to power the three key areas of 'communications, learning and knowing and research' (Snowden 2003). Whether it's a couple of words or pages long, or a trend in a graph, narratives are arguably the most knowledge-dense of our communications with each other.

In a forum of the American Psychological Society, de los Reyes (2024) observes that facts without stories turn people off, which explains in a succinct manner why I have used a narrative approach. Furthermore, the reviewer of a major book on tropical diseases once remarked that describing the shape of a fungal spore as 'ellipsoidal, obovoid, spherical or subspherical' would not help the reader as much as a simple line drawing. Pictures and the heuristics that stories often contain are critical for timely and precise disease diagnosis in the field.

This book is therefore intended to be neither didactic nor encyclopaedic, but collaborative and interactive. It concerns my personal experiences and perspectives from field visits to crops, growers and countries, presented in a largely narrative and visual way. This narrative approach serves different functions for its three groups of audiences.

The first group are professional scientists and specialists. By engaging in the narratives of others, or by sharing their own, the reader applies his/her own knowledge and experiences. In doing so, new perspectives/insights that are unique to each individual often emerge. Oftentimes it may even involve the recollection of the individual's own 'long-forgotten stories'. Some knowledge management circles call this helping people 'know what they did not know they knew'. This interactive/collaborative approach is independent of whether the reader agrees with the opinions of the author, which is the essence of scientific discourse.

For students, it serves the function of maximizing the transmission of tacit, experiential knowledge, rather than simply providing yet another set of reference data destined for library shelves. My students at HAFL (Bern University of Applied Sciences) tell me that what they appreciate most from teachers is not more facts, but experiential knowledge. Recent progress in Generative AI suggests that the future role of textbooks may lie in facilitating learning, rather than serving just as repositories of facts (see Chap. 8).

Thirdly, it seeks to attract the non-specialist reader with interests in topical areas like sustainability, biodiversity, food security, climate change and investment in new technologies, by offering an easily readable text and concise information about the relevance of tropical crops.

This book aims to meet all these objectives whilst maintaining the rigour of a fully referenced report of scientific work.

The underlying theory behind these three practical considerations of the uses of story also comes from complexity science. Much of scientific investigation today is still undertaken to find out '*what we know we do not know*'. If we are to address the complexity of issues in the future, however, we need to shift our attention towards knowing what '*we do not not know we do not know*' (i.e. the area of complexity). Lessons from the past in human affairs and in science tell us that the most important 'black swans' and perhaps the best discoveries of the future will lie in unexpected areas. Engaging with the stories of others is one way to help us gain new perspectives that we may have previously been unaware of.

To gain further perspective, these brief narratives (partly in sections I call 'Storypads', and partly as snippets) are complemented by bite-sized chunks of statistical data, mainly sourced from the FAOStat database (https://www.fao.org/faostat/en/#home) and CGIAR Institute (https://www.cgiar.org/) publications. Two extensive appendices in a dashboard-like format provide rapid access to referenced data on leading disease control agents (chemical and biological) and an overview of all control measures (including genetics, cultural and control agents) for all diseases mentioned in the text. Essentially, this provides the 'heavy part' of the book to complement the lightness of the tales.

1.4 How Is the Book Organized?

A chapter on each crop will therefore describe its significance and geographical distribution, followed by a panoramic view of the damage caused by the major diseases with a pictorial story of my own impressions and experiences. The latter may include an itinerary, as well as related points of interest during the visit. A minimum of technical terms will be used, although for precision and reference, Latin names and systematics are always applied.

To bring together the chapters on different crop groups, I have included two concluding chapters. The 'Future: food security, sustainability and resilience' puts into context the implications of past experiences for a future driven by 'freight train trends' (Martin 2008). It also suggests new approaches to the study of climate change based on complexity theory. The 'Epilogue' weaves the stories together into a single narrative and suggests ways to use the book.

In keeping with scientific tradition, comments are evidence-based or appropriately referenced unless otherwise qualified. Crops and groupings selected for discussion are mainly those in which I have developed a personal interest. For the sake of completeness, I also added, on occasion, 'Other Crops' to each grouping, those with which I have only had a passing acquaintance. To make up for the brevity of treatment, an attempt will be made to point out in '*Extended Reading*' some interesting sources to which the reader could go for further information. References contain approximately 240 citations, mainly from published sources. The appendices provide quick access to a library of facts as described above.

A text I commend to the reader for further reading is that by Burkill (1935, updated 1966), who was a former Director of the Botanical Gardens in Singapore. 'The dictionary of the economic products of the Malay Peninsula' is a scholarly text of over 2400 pages covering Malaysia, India and some other countries in SE Asia. It is not only informative but an enjoyable read with fascinating local details of how each plant is locally used, written by an author with an intimate, personal experience of the region.

The pictorial materials I shall share in this book are entirely my own and were largely used before to illustrate a degree course on Tropical Plant Pathology at the Bern University of Applied Sciences (HAFL, Zollikofen) from 1999 to 2007. Interestingly some modern postgraduate courses at HAFL today have taken an approach of self-learning similar to that I describe (see Chap. 8 on the Future).

The book does not try to duplicate several excellent texts available today which comprehensively describe specific crop diseases in the tropics (e.g. Ploetz 2005, 2006), or use of narratives in a different way on specific sets of crops (e.g. Schumann and D'Arcy 2017).

References

Alvarez S, Rohrig E et al (2016) Citrus greening disease (Huanglongbing) in Florida: economic impact, management and the potential for biological control. Agric Res 5(2):109–118

Bruns HA (2017) Southern corn leaf blight: a story worth retelling. Agron J 109:1–7

Burkill IH (1966) A dictionary of the products of the Malay peninsula. Ministry of Agriculture and Cooperatives, Kuala Lumpur, p 2444

Colussi J, Schnitkey G (2021a) New soybean record: historical growing of production in Brazil. Farmdoc daily 11:49

Colussi J, Schnitkey G (2021b) Brazil: Corn Production in Three Crops per Year. Farmdoc daily 11, 58, Department of Agricultural and Consumer Economics, University of Illinois at Urbana-Champaign

De Los Reyes A (2024) let me help you tell a story: strategies for improving writing clarity. The Journal of Clinical Child and Adolescent Psychology Future Directions Forum. American Psychological Association. https://www.apa.org/career-development/writing-clarity.pdf

De Souza JT, Monteiro FP, Ferreira MA, Gramacho KP, Martins ED, Luz N (2018) Cocoa diseases: witches' broom. In: Umaharan P (ed) Achieving sustainable cultivation of cocoa. Burleigh Dodds Science Publishing, Cambridge

Dickinson M (2015) Loop-mediated isothermal amplification (LAMP) for detection of Phytoplasmas in the field. In: Lacomme C (ed) Plant pathology. Methods in molecular biology, vol 1302. Humana Press, New York, p 99. https://doi.org/10.1007/978-1-4939-2620-6_8

Dos Santos GC, Lima Horn LM, Trezzi Casa R, Soardi K, Lopes MA, Nascimento SCD, Santi VM et al (2024) First report of seed decay caused by Diaporthe ueckeri on soybean in Brazil. Plant Dis. https://doi.org/10.1094/PDIS-04-24-0814-PDN

FAO, IFAD, UNICEF, WFP and WHO (2021) The State of Food Security and Nutrition in the World (SOFI). Transforming food systems for food security, improved nutrition and affordable healthy diets for all. Rome, FAO. https://doi.org/10.4060/cb4474en

References

Gilbertson RL, Macedo MA, Maliano MR, Rojas MR (2021) Weed-infecting viruses in a tropical agroecosystem presents different threats to crops and evolutionary histories. PLoS One 16(4):e0250066. https://doi.org/10.1371/journal.pone.0250066

Gurr GM, Johnson AC, Ash GJ, Wilson BAL, Ero MM, Pilotti CA, Dewhurst CF, You MS (2016) Coconut lethal yellowing diseases: a *Phytoplasma* threat to palms of global economic and social significance. Front Plant Sci 7:1521. https://doi.org/10.3389/fpls.2016.01521

Harlan JR (1976) The plants and animals that nourish man. Sci Am 235(3):88–97

Kessmann H, Staub T, Hofmann C, Maetzke T, Herzog J, Ward E, Uknes S, Ryals J (1994) Induction of systemic acquired disease resistance in plants by chemicals. Annu Rev Phytopathol 32:439–459

Liberei R (2007) South American leaf blight of the rubber tree (*Hevea* spp.): new steps in plant domestication using physiological features and molecular markers. Ann Bot 100(6):1125–1142. https://doi.org/10.1093/aob/mcm133

Martin J (2008) Target Earth: the grand scale problems of the 21st Century, Public Lectures of the Oxford Martin School. University of Oxford. https://www.oxfordmartin.ox.ac.uk/events/public-lecture-target-earth-by-james-martin

Mena E, Stewart S, Montesano M, de Leon IP (2023) Plant Pathol 73(1):31–46. https://doi.org/10.1111/ppa.13803

Milne AE, Teiken C, Deledalle F, van den Bosch F, Gottwald TR, McRoberts N (2018) Growers' risk perception and trust in control options for huanglongbing citrus-disease in Florida and California. Crop Prot 114:177–186

Mpunami AA, Tymon A, Jones P, Dickinson M (1999) Genetic diversity in the coconut lethal yellowing disease phytoplasmas of East Africa. Plant Pathol 48(1):109–114

Ou SH (1985) Rice diseases, 2nd edn. Commonwealth Mycological Institute, Kew, p 380

Padmanabhan SY (1973) The great Bengal famine. Ann Rev Phytopathol 11:11–26

Ploetz RC (2005) Panama disease, an old nemesis rears its ugly head: part 1, the beginnings of the banana export trade. Plant Health Progress 6. https://doi.org/10.1094/PHP-2005-1221-01-RV

Ploetz RC (2006) Panama disease, an old nemesis rears its ugly head: part 2, the cavendish era and beyond. Plant Health Progress. https://doi.org/10.1094/PHP-2006-0308-01-RV

Santini A, Liebhold A, Migliorini D (2018) Tracing the role of human civilization in the globalization of plant pathogens. ISME J 12:647–652. https://doi.org/10.1038/s41396-017-0013-9

Schneider K, van der Werf W, Cendoya M, Mourits M, Navas-Cortés JA, Vicent A, Lansink AO (2020) Impact of *Xylella fastidiosa* subspecies *pauca* in European olives. PNAS 117(17):9250–9259. https://doi.org/10.1073/pnas.1912206117

Schumann GL, D'Arcy CJ (2017) Hungry planet: stories of plant diseases. APS Publication

Schumann GL, Leonard KJ (2001) Stem rust of wheat (black rust). Plant Health Instructor. https://doi.org/10.1094/9780890544907

Snowden DJ (2003) Narrative patterns: the perils and possibilities of using story in organisations. In: Lesser E, Prusak L (eds) Creating value with knowledge. Oxford University Press, p 14

Talhinhas P, Baptista D, Innes DINI, Vieira A, Silva DN, Loureiro A et al (2017) The coffee leaf rust pathogen *Hemileia vastatrix*: one and a half centuries around the tropics. Mol Plant Pathol 18(8):1039–1051

Webb JLA (2002) Tropical pioneers: human agency and ecological change in the highlands of Sri Lanka. Ohio University Press, Athens

Chapter 2
Time to Check Our Bearings Before We Start On Our Journey

Abstract Since our journey will be largely based on my personal interests and experiences, it would perhaps be useful to ask how well these personal anecdotes reflect trends in the 'real world'.

Data from about 600 science documents on tropical crop diseases were extracted from CAB abstracts and Biosis over the last 65 years, represented by three arbitrary 5-year periods: 1957–1962, 1987–1992 and 2017–2022. Using text mining procedures, statistics on crops and diseases were extracted from the abstracts, and the trends are plotted in the following graphs (respectively Figs. 2.1 and 2.2).

2.1 Crop and Disease Trends Over the Period 1957 to 2022

The crops and diseases contained in this analysis do not necessarily represent those of the greatest economic importance but are those that have attracted the highest priorities for research. Since these could have been published for a variety of reasons, including impact (global and local), funding, suitability of research topics for academic study, availability of technology and so forth, one can only assume that they are thought to be of the greatest benefit to agriculture and society. It is nevertheless reassuring that the crops and diseases we shall cover in this book are, in general, in agreement with those prioritized in the literature.

In keeping with the spirit of this book, the second reason to share these results is that it should facilitate self-learning.

In this case, the reader is invited to 'join the narrative', e.g. by questioning themselves, which trends make sense, what do they mean, and what new perspectives and insights do they stimulate (see Crop and Disease trends for elaboration), based on their own experiences and knowledge. In this manner, it is likely that new insights unique to each reader could emerge.

By adding author and institutional fields, these trends can also help create new research networks, although that's a question that goes beyond the scope of this book.

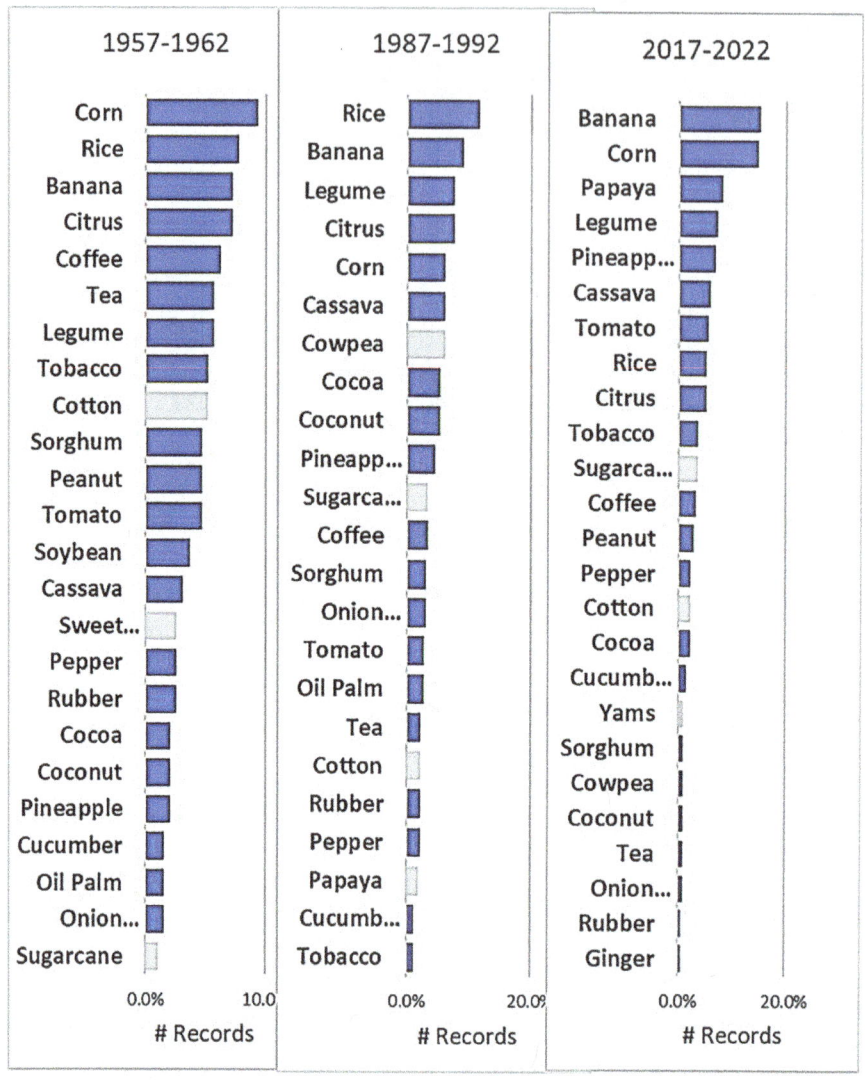

Fig. 2.1 Crop trends over the last 65 years based on tropical crop disease citations (see text for analytical details). Filled boxes refer to crops mentioned in text; empty boxes, not mentioned

2.2 Tales from Some of the Crop and Disease Trends

To set the scene, I have listed here some perhaps obvious examples.

2.2 Tales from Some of the Crop and Disease Trends

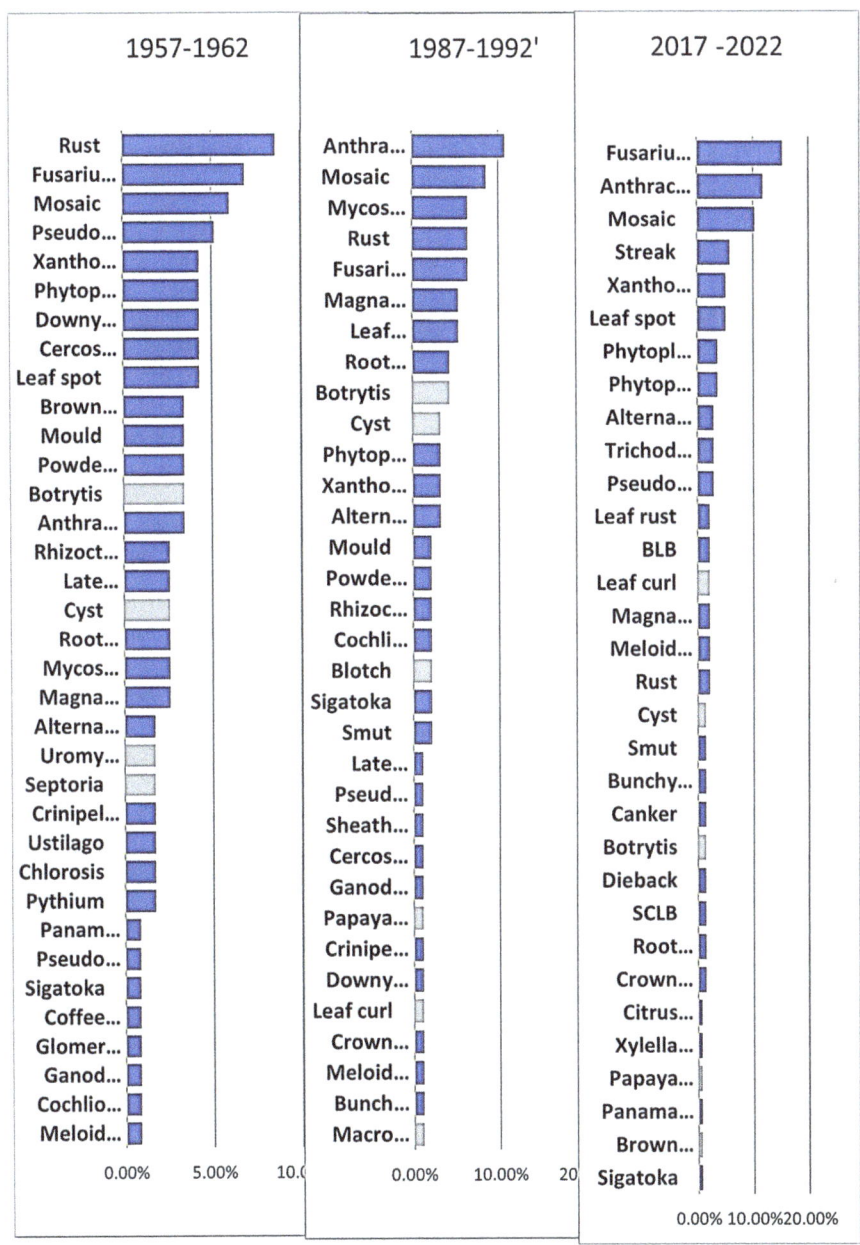

Fig. 2.2 Disease trends over the last 65 years based on tropical crop disease citations (see text for analytical details). Filled boxes refer to diseases mentioned in text; empty boxes, not mentioned

2.2.1 Crop Trends (see Fig. 2.1)

The rice trend: Over the three survey periods, the percentage of documents about rice and its diseases was respectively 15% (1957–1962), 31% (1987–1992) and

11% (2017–2022). This high proportion of rice-related reports may partly be attributed to the impact of IRRI. IRRI was established in 1960 and sparked not only IR8 and the green revolution in rice production during the late 1960s–1970s, but an unprecedented focus on rice research and training all over the rice-growing world over the next 50 years. This included a large amount of work on pests and diseases to protect the higher yields that the miracle rice could achieve. In recent years, the threat to food security has placed the emphasis again on increasing yields (e.g. with hybrid rice), so that the proportion of publications on disease has relatively decreased.

The banana and fruits trend: There are perhaps three reasons why bananas top the research trends. The continuing threat of Panama disease over the last 60–70 years (the pathotype TR4 continues till today, spreading to new geographic areas in Africa and Australasia; see Storypad 6); the need for control of the Sigatoka diseases (not only for maintaining high yields, but because low amounts of infection can impact the transport quality of fruits over their long sea journeys); the threat of epidemics over large, contiguous areas with uniform genotypes; and the problems arising from micropropagation.

Besides bananas, the other major tropical fruits mango, pineapple, avocado and papaya have been amongst the world's fastest-growing agricultural commodities (FAO 2020). It is scarcely surprising to see pineapple and papaya rapidly climbing the charts over the three survey periods.

Cassava is another crop that has been climbing the publication charts from 6%, 16%, to 13%. One of man's oldest subsistence crops, it tops food security needs because of its low nutrient and input requirements, resistance to pests and ability to survive fires and conflict. Is its trending a sign of troubled climatic and social times? Attention has also been directed recently to virus resistance.

Legumes are trending, perhaps to meet the sustainability needs of today by producing their own requirements for nitrogen fertilizers, as well as for other crops when used as cover or rotation crops in sustainable agriculture, whilst meeting increasing protein needs in food.

The spectacular increase in the production of soybeans in Brazil noted in Chap. 1 is expected to increase further in the next decade, requiring greater yield intensification and protection against pests and diseases to improve sustainability.

2.2.2 Disease Trends (Fig. 2.2)

The consistent high ranking of Fusarium wilt (including Panama disease) may be largely linked to its impact on bananas and cucurbits.

The increase in mosaic and streak diseases is largely due to trends in virus diseases of cassava (see Crops). Anthracnose and Phytophthora have been trending because of the wide range of crops they attack.

Phytoplasmas have gained in recent importance because of their impact on sugarcane and coconuts. Xanthomonas has risen in the charts linked to an increase in bacterial blight on cassava.

Without labouring the point, a virtual treasure trove of tales can be extracted from these trends.

Ask yourself, as you browse through the trends, it may be useful to consider if they enhance or detract from your previous perspectives. In this regard, focusing on unexpected trends may be more productive than those that are expected.

We now have our bearings because we can see where the crops and diseases we shall visit sit in the overall context of tropical plant pathology over the last 65 years (viewed from an analysis of scientific literature). It is perhaps reassuring that most of the crops and diseases referred to during our journey also appear in this random survey (see Figs. 2.1 and 2.2).

2.2.3 Tale of the Ubiquitous 3 in the Tropics

Three pathogens stand out as being almost ubiquitous on tropical crops. For example, anthracnose (caused by Colletotrichum) occurs on avocado, banana, beans, citrus, guava, mango, mangosteens, papaya, pepper, sweet potatoes and wax apple; Phytophthora rots and blights occur on aubergine, avocado, black pepper, citrus, cocoa, coconut, durian, papaya, pineapple, taro and tomato; and Fusarium wilts on banana, carrots, coconut, cucumber, maize, oil palm, onion, potato, rice, soybeans, tomatoes and sweet potato.

Two other pathogens are also common. Alternaria is known to infect beans, cabbage, carrots, cucumbers, onions, pepper and potato; and Rhizoctonia attacks aubergines, beans, maize, rice, soybean and tomato.

Clearly, these organisms have enjoyed significant evolutionary success in adapting to and colonizing a diverse range of hosts in the tropics. The physiological and genetic reasons for their adaptability remain to be fully explored.

2.2.4 A Gentle Reminder

Throughout this journey, the reader is encouraged to reflect on whether he or she strongly agrees or disagrees with specific tales and the new personal perspectives that may arise therein (see also Chap. 9, Epilogue).

Reference

FAO (2020). Medium-term Outlook: Prospects for global production and trade in bananas and tropical fruits 2019 to 2028. Rome.

Chapter 3
Cereals

Abstract We start our journey with cereals because these crops produce more food energy (51% of calories and 47% of protein in the average annual diet) than any other crop group (World Bank WDI database 2000 to 2020, based on FAO data). Maize, wheat and rice are the world's most widely grown cereals. Maize leads with almost 1100 million tonnes harvested in 2018/2019, followed by wheat (734 million tonnes) and rice (495 million tonnes, FAOSTAT, 2020). Together, these three cereals account for nearly 90% of world cereal production. Maize is grown mostly, and rice almost exclusively, in warm temperate to tropical regions.

3.1 Rice

Extended reading: Ou (1985), Cartwright et al. (2018), IRRI (2024)

With the reader's indulgence, I shall dwell a little longer on rice than on other crops because I spent much of my formative years in plant pathology working on it before moving on to temperate crops.

3.1.1 Production

Rice is the staple diet for about half the world's population and provides some 21% of the total global calorie intake (Yuan et al. 2021) but uses only 11% of the global cropland. Archaeological evidence suggests that rice cultivation began in the coastal wetlands of eastern China at least 8000 years ago (IRRI 2007). It is grown today in over 100 countries, mainly in tropical regions, but also in subtropical and warm temperate areas with a frost-free period of over 130 days.

Rice is mainly cultivated under irrigated conditions with surface water. A total of 79 million ha of irrigated lowland rice provide 75% of the total world production. The remaining 25% is rainfed or upland. In Latin America (e.g. Brazil, Colombia), however, most rice is upland and mechanized.

3.1.2 Storypad 2: 'Tall Tales' of Rice: The Astonishing Diversity

*Traditional varieties of rice in Asia are often tall (**120–150 cm**), but weak-stemmed and respond to fertilizer application by lodging, which in turn reduces yield. Others, like Dee-geo-woo-gen, are dwarfs, typically about **60 cm** in height. Beginning with 'IR8' in the seventies, the new 'miracle' rice breeds from the International Rice Research Institute (IRRI 2016) that were driving the 'green revolution', were semi-dwarfs (**110-120 cm** tall). They had stiff straw, were highly yield-responsive to fertilizers, and improved rice production across the world wherever irrigation could be controlled. New hybrid cultivars can be **2 m** tall (see later).*

*On the other hand, deep water rice grows under conditions of more than 50 cm of uncontrolled flooding for at least a month. Additionally, floating rice varieties may need to cope with water depths of up to **4–5 m** (e.g. in the Chao Phraya delta region of Thailand). These cultivars survive by stem and leaf elongation, keeping pace with rising floodwaters, and kneeing (bending upwards) of the terminal parts of the plant to keep the panicle above water as the floods subside (Puckridge et al., 2000). Often, where irrigation is available, the flooded rice is followed by a rotation of upland field crops in the dry season (e.g. Vietnam).*

Hybrid rice technologies pioneered by Longping Yuan in China (since the 1960s) are known to be capable of outperforming the best inbred rice but have previously been held back because of the lack of a practical system for the production of hybrid rice seed. The problem has now been resolved using MS restorer lines in collaboration with IRRI. Due to heterosis (hybrid vigour), plants are larger (2 m tall!) and can sustain yield increases of 30–40% (IRRI 2007). These cultivars also show multiple resistance to diseases, pests and environmental stresses, and are already grown on an estimated 50% of China's rice area.

3.1.3 Itinerary

- Thailand: One of the world's largest exporters of rice. Main rice areas are in the north-east and the central plains. Twenty-five percent is irrigated; the rest is rain-fed. The Ayuthaya region in the Central Plains is also home to deep water and floating rice cultivation.
- Republic of Indonesia: A leading producer *and* consumer (185 kg rice/capita/year). Most rice is grown on the island of Java, mainly in irrigated fields.
- Japan: Small farms (less than 1 ha) and highly mechanized. Consumes 73kg/capita/year.
- Colombia: Second-largest rice producer in Latin America after Brazil. Largely upland and mechanized.

3.1 Rice

- Italy: Largest producer in Europe. Mainly irrigated 'risotto' (japonica) rice in the Lombardy and Piedmont regions. Probably the smallest per capita consumer among producing countries, at 9 kg/capita/year in 2023
- Consumption per capita provides an indication of the importance of the crop for food security. (data extracted from FAOStat 2023)

3.1.4 Rice Disease Panorama

Rice Blast Disease

Rice blast, caused by *Magnaporthe oryzae* (synonym *Pyricularia oryzae*), is probably the most widespread and destructive disease of rice (Fig. 3.2). It was one of the earliest diseases of rice to be described in the literature. Ou (1985) tells the story about the 'rice fever disease' being already reported in 1637 (late Ming dynasty) in a publication on Natural Resources in China).

Under conducive conditions, the attack on the foliage (leaf blast) can stunt or kill whole plants before they mature (Fig. 3.2d). As might be expected from the common name plants can look like they have been subjected to a blast of hot air. Moderate attack on foliage can cause a yield loss of 30–35%. The second phase of the disease causes significantly higher losses due to the attack at the nodal or neck region, or other parts of the panicle. Early infection reduces grain filling, resulting

Fig. 3.1 Starting off. (**a**) Indonesia: manually transplanted rice, flooded, typical of much of Asia. (**b**) Colombia: upland highly mechanized cultivation. (**c**) Italy: direct seeded, intermittent flooding. (**d**) Japan: seedling nursery for mechanical transplanting. (**bk**) Thailand: setting off by boat to visit deep water rice

Fig. 3.2 Blast disease of rice. (**a**) Conidium of *Magnaporthe oryzae*, the causal fungus of the blast disease of rice. (**b**) Elliptical/diamond-shaped lesions of leaf blast disease. (**c**) Neck rot symptom of blast on rice panicle. Note severance of ear at basal node under the weight of the panicle when infection comes after the grains have filled. All parts of the panicle can be attacked and covered by a bluish-grey fungal growth (Chin 2005). Panicles are empty, bleached and stand upright if infection occurs before grain filling. (**d**) Field symptoms of burnt leaves due to blast disease, as if they had been subjected to a 'blast' of hot air
NB: see 3.1.6 for comparison with damage due to insects

in empty heads. Later, 'neck rot' (Fig. 3.2c) often results in the detachment of the filled panicle under its own weight (CPC datasheet, 2005, on *Magnaporthe oryzae*).

The disease can be found virtually wherever rice is grown, in Africa, Asia, Oceania, North and South America and Europe. Plants subjected to long or frequent periods of rain or high humidity are generally more at risk; so are plants grown under moisture stress in upland cultures. Controlled field trials have produced yield losses of 21–42% (Chin 1975).

More recently, similar symptoms of elliptical lesions have caused concern on wheat (first in Brazil in 1985, and subsequently in the US) (Cruz et al. 2016). Wheat blast is caused by a host-specific pathotype of *Magnaporthe oryzae Triticum* (MoT).

The disease is best managed through the use of resistant cultivars (as well as cultivar mixtures, Chin and Ajimilah 1982; Zhu et al. 2000), the application of fungicides and minimizing the use of excessive nitrogenous fertilizers (see last chapter on the future and appendices).

The pathogen has, however, displayed the ability to adapt genetically to both resistance genes and fungicides. Host resistance to leaf blast is not always correlated with panicle blast resistance. The classical pyramiding of R-genes in host cultivars is one solution, but this can reduce crop growth and increase linkage to undesirable characteristics. Molecular approaches are being explored to address these challenges (Ning et al. 2020).

Sheath Blight Disease

Sheath blight of rice is caused by *Rhizoctonia solani.* (teleomorph *Thanatephorus cucumeris*) (Fig. 3.3). We shall encounter this ubiquitous pathogen again on other crops later in the book (e.g. see soybean, corn, tomato and aubergines), as well as several grass weeds. It is perhaps, therefore, not surprising that no significant levels of host resistance have been found in any variety of rice.

As an interesting adaptation to an aquatic environment, the disease is spread mainly through sclerotia (Fig. 3.3c), aggregations of fungal mycelium designed to enable survival in the soil or on crop debris between crops, which float and disperse on the surface of irrigation water.

The disease is estimated to cause about a 6% yield loss of rice in the lowland tropics (IRRI). Under conducive conditions, field trials have produced yield losses of 31–44% (Chin 1977). A range of effective fungicides has been known since the late 70 s (see Appendix), but cultural methods and thermosanitation may be more economical in much of the developing world.

Fig. 3.3 Sheath blight disease of rice caused by *Thanatephorus cucumeris* (imperfect state, Rhizoctonia solani). (**a**) Lesions on leaf sheaths. (**b**) Basidiospores of *T. cucumeris* (causing sheath blight of rice), looking like the four points of a crown. These are thought to be responsible for the aerial spread of the disease. (**c**) The 'batik' design of sheath blight disease on the leaves (and leaf sheaths, see above) of rice. Light brown coloured sclerotia are visible at the lower end of a leaf blade. These resting bodies enable the survival of the fungus between crops and are able to spread the disease by floating on the surface of irrigation water

3.1.5 Storypad 3: Sheath Blight Disease (the 'Painter Who Would Be King')

The colourful symptoms on the leaves and stems (Fig. 3.3) resemble batik (a popular wax-print of fabric in Indonesia and Malaysia). The causal organism is Rhizoctonia solani (Chin and Supaad 1986).

The sexual state (teleomorph) of the fungus (Thanatephorus cucumeris) appears totally different, resembling toothpaste foam on green, apparently healthy stems. Magnified, the regal, crown-like basidia (fruiting bodies) appear, each carrying four basidiospores (Fig. 3.3b) at the points of the crown.

In nature, the sexual state (teleomorph) of the fungus on rice is relatively rare in occurrence. It was first reported in Malaysia only in 1976 (Chin 1976). Perhaps this is not entirely surprising, since the teleomorphs of 20% of the imperfect fungi are as yet unknown (Dyer andd Kück 2017). Mycelial cells of R. solani are multinucleate and can generate genetic diversity through the well-known parasexual cycle common among imperfect fungi. It does not produce resilient fruiting bodies for the purposes of survival. So the question may be raised, how does the fungus benefit from having a teleomorph? Perhaps part of the reason could be that aerial spread by basidiospores is likely, as evidenced by the occasional occurrence of small isolated lesions located on leaves high up on a plant in the field.

Leaf Spot Diseases

Brown leaf spot was the disease thought to have contributed to the disastrous Bengal famine in 1943. The causal pathogen is *Cochliobolus miyabeanus*.

The main symptoms are the large oval brown spots on the leaves, which turn grey with age (Fig. 3.4a). Its cigar shaped spores are shown in Fig. 3.4b. The disease also attacks the ears and grains (Fig. 3.4c) and is a major component of the complex 'dirty panicle disease'. Other components of the complex are *Alternaria padwicki* and *Curvularia* spp. Controlled field trials of brown spot have shown yield losses of 8–12% due to infection (Chin 1974).

Another common leaf disease that is self-descriptive is narrow brown spot, caused by *Cercospora oryzae (syn Cercospora janseana)*. It is important in the southern states of the United States and is common but not generally considered a major disease in Asia.

Cochliobolus and *Cercospora* are key components of the complex of fungi that cause the dirty panicle disease of rice. Other fungi commonly observed include *Alternaria, Curvularia and Heterosporium*.

False and Kernel Smut Diseases

Wearing a white shirt when entering a field containing false smut is unwise, as one is likely to emerge with one that is stained yellow. The disease replaces grains with large balls of yellow spores (Fig. 3.6b).

Besides a loss in yield and the increased chalkiness of grains, there is a potential for mycotoxins to occur in the harvest. Seed treatment, resistant cultivars and field

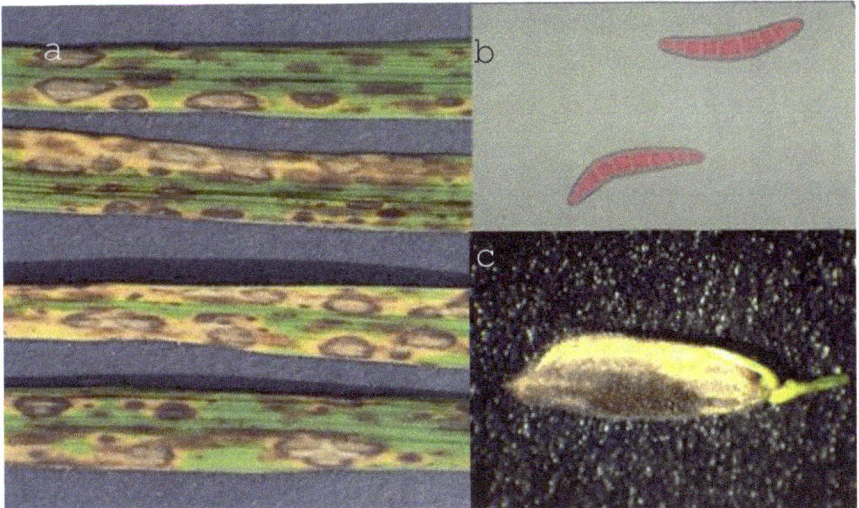

Fig. 3.4 Brown spot disease of rice caused by *Bipolaris oryzae* (teleomorph *Cochliobolus miyabeanus*). (**a**) Ovate brown spot lesions on rice leaves. (**b**) Cigar-shaped conidia of *Bipolaris oryzae*. (**c**) Infected grain covered by conidia. This is a common symptom of the dirty panicle disease, which is caused by a complex of different fungi

sanitation have been reported to be effective in controlling the disease (Song et al. 2021; Chi et al. 2020).

Kernel smut, on the other hand, replaces the grain with a ball of dark spores (Fig. 3.6a).

Stem Rot

Stem rot is caused by *Magnaporthe salvinia*. The leaf sheaths and stems of the rice plant are sequentially attacked, resulting in lodging or reduced yields. The disease is spread mainly by sclerotia (dry masses of fungal mycelia, Fig. 3.5b), which can remain viable in soil or crop debris for up to 6 years.

Unlike sheath blight sclerotia which spread on the surface of irrigation water, these sclerotia are largely plant debris/soil-borne. Several other sclerotial diseases of rice have been reported by Chin and Supaad (1986).

Bacterial Leaf Blight and Bacterial Leaf Streak Diseases

These bacterial diseases are caused, respectively, by *Xanthomonas campestris* pv *oryzae* and *X. campestris* pv *oryzicola*.

Leaf blight (Fig. 3.5a) appears initially as water-soaked lesions, later turning into white wavy stripes along leaf margins, giving the plant a scorched appearance (Chin and Supaad 1986). Leaf streak produces narrow linear lesions appearing yellow and translucent when held against the light. The entire leaf dies from coalescing lesions.

Yield losses of about 10% have been recorded due to leaf blight.

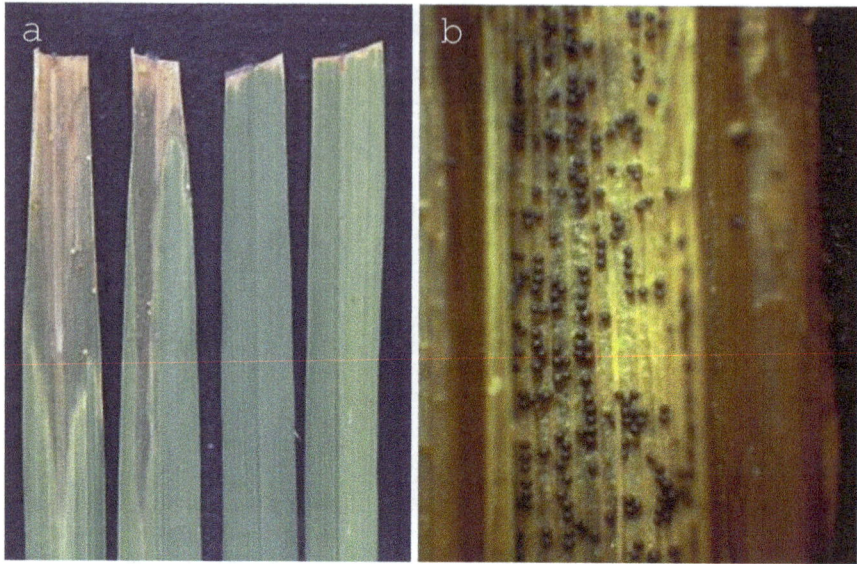

Fig. 3.5 Bacterial leaf blight disease of rice caused by *Xanthomonas campestris* pv *oryzae* and stem rot caused by *Magnaporthe salvinii*. (**a**) Water-soaked stripe lesions of bacterial leaf blight on leaves. Note bacterial ooze emerging from lesions. (**b**) Stem rot of rice caused by *Magnaporthe salvinii* (previously *Helminthosporium sigmoideum*). Stem split to show sclerotia (resting body formed from compacted mycelium) of fungus. These enable survival in the soil and in crop residues

The use of resistant cultivars has been the most important method of control. Recent use of molecular markers has improved breeding for resistance (Chukwu et al. 2019).

Virus Diseases

Rice tungro disease is caused by two viruses (RTSV/*Waikavirus* and RTBV/*Tungrovirus*), transmitted by leaf hopper insects (*Nepthotettix virescens*, Fig. 3.7 inset). Infected plants are severely stunted, tillering is little affected, and leaves are discoloured reddish or orange (Fig. 3.7), hence Penyakit Merah or red disease in Malaysia (Chin and Supaad 1986). Severe infection can result in almost total loss of yield.

The disease can be detected by iodine reaction due to starch accumulation in the leaves, ELISA methods, or more recently by PCR tests of either hoppers or plants. Resistance to the disease is largely due to vector resistance rather than virus resistance.

3.1.6 Tales of Mistaken Identity

Brown plant hopper damage on rice can look, at a distance, to be similar in appearance to damage caused by leaf blast. High populations of hoppers desiccate whole plants, causing 'hopper burn', but can obviously be differentiated by the absence of

3.1 Rice

Fig. 3.6 Seed-borne diseases of rice. Dirty panicle disease is caused by a fungal complex often made up of *Cochliobolus, Cercospora, Alternaria, Curvularia, Heterosporium* spp. (**a**) Kernel smut of rice caused by *Tilletia barclayana* (insets showing chlamydospores and their germination to form sporidia). Grains are replaced by fungal growth, which bursts through at maturity to form charcoal-black spore masses. (**b**) False smut of rice caused by *Ustilginoidea virens*. Conspicuous, large yellow to orange spore balls replace the grains. (**c**) Stackburn disease caused by *Alternaria padwickii* (inset) on grain as well as on leaves (not shown)

blast lesions and the presence of insects.. Stem borer damage can be similar to stem rot disease (see above) or empty panicles caused by panicle blast disease. Like tungro, rice ragged stunt also causes stunting of the rice plant but is accompanied by ragged and twisted leaves and incomplete emergence of the panicle (Habibuddin 1978). On the other hand, rice bakanae disease (aka the 'foolish disease') is caused by the fungus Giberella fujikuroi. The giberellin produced by the fungus causes the shoots of the rice plant to elongate and become paler green in colour (Chin and Supaad 1986).

Common damage due to environmental extremes includes those caused by saline or acid sulphate conditions of the soil can sometimes be similar to viral disease damage.

3.1.7 Rice Disease Control

See Appendices for details

The most common method of disease control, which has been actively promoted by CGIAR institutes like IRRI, is the use of resistant cultivars of rice. These may be deployed as single cultivars, in cultivar mixtures, or as genetic elements of hybrid

Fig. 3.7 Tungro virus disease of rice is caused by a composite of the two viruses RTSV (*Waikavirus*) and RTBV (*Tungrovirus*). The viruses are transmitted by leaf hoppers (*Nephotettix virescens and nigropictus*, inset). Philippines. Infected plants are stunted with orange to reddish discoloration of the leaves. Losses can range from 46% to 100% (Lim 1972; Habbibuddin, pers. comm)

rice. Cultural methods, like minimizing the use of excessive nitrogenous fertilizers, the application of supplements like basic slag, thermosanitary methods with infected stubble and composting of straw, may be used to help reduce disease pressure.

Most agronomic measures, ranging from the timing of irrigation, planting or transplanting density, plant nutrition and sanitation practices, contribute to the integrated management of pests and diseases.

Pesticides are widely used within the guidelines of integrated pest management programs.

Both resistance genes and fungicides are, however, subject to genetic adaptation by pathogen populations (e.g. especially those of *Magnaporthe oryzae*). Use strategies to prolong their durability have therefore been developed based on the modelling of selection and fitness in pathogen populations and field testing (Chin 1986). Recent approaches have used molecular breeding, which includes R genes, defense-regulator genes and quantitative trait loci (QTL), to deliver more broad-spectrum and durable sources of resistance (Liu et al. 2021).

3.1.8 Rice Disease Pictures

Composite pictures in this book are always labelled in clockwise order from top left as: *a, b, c, d* or *bk* (background).

3.2 Maize (Corn)

Maize was first domesticated in Mexico and introduced by the native Indians (who called it maize) to European settlers in North America.

3.2.1 Production

The top producers in 2020 were the United States (360 million tonnes), China, Brazil, Argentina, Ukraine, Mexico and India (FAOSTAT, 2022). The 13 corn belt states in the United States alone produced about a third of the world's production from 1994 to 2020.

Corn, as the crop is referred to in the United States, is grown mainly for food and food-related products, animal feed (about a third of production in the United States) and industrial uses including ethanol production. Major specialty items include dextrose and refined oil (food and drugs), dextrins (industrial), corn starch (food, drug, cosmetics, industrial), corn syrup and corn starch (food and industrial) (USDA, Economic Research Service; *Texas Farm Bureau, Research and Education Department*).

Maize yields in the tropics are typically lower (about 46% of those in temperate regions) but are increasing in both regions at about the same rate (70–80 kg/ha/year). In sub-Saharan Africa, the crop is the staple food, with 95% of the crop being used directly for human consumption (Kumar et al. 2014b).

Produced mainly in Bangladesh, China, India, Indonesia and Pakistan, maize is now the second most important food and feed crop after rice in Asia. Although its use as food is decreasing, it is the primary source of feed for the poultry and livestock industry as well as a source of raw material for the manufacturing sector. It is therefore an important source of livelihood and food security for many Asian families (Kumar et al. 2014a).

3.2.2 Maize Disease Panorama

See Appendices for disease control measures

A wide range of bacterial, fungal and viral diseases occur on the crop. Different sets of diseases appear to be adapted to the tropical and temperate regions, perhaps in response to the range of environments under which the crop is grown. Aflatoxin exposure due to *Aspergillus* infection/contamination can also be important under conditions of subsistence farming in the tropics, wherever storage conditions are suboptimal (see aflatoxins in Field Crops: Groundnuts).

Renfro and Ullstrup (1976) observed that the distribution of maize diseases tends to be temperature-dependent. Warm temperate to tropical regions are preferred by late wilt, banded leaf, sheath blight, *Rhizoctonia, Botryodiplodia,* grey ear rots, *Sclerospora*-incited downy mildew diseases, southern rust, *Pythium* and bacterial stalk rots, *Curvularia* leaf spot and ear rot caused by *Diplodia macrospora.*

On the other hand, bacterial wilt, eye spot, yellow leaf blight, northern leaf blight, common rust, as well as stalk and ear rots caused by *Diplodia maydis* and *Gibberella zeae* are usually restricted to temperate areas.

Recently, a newer disease, Goss's bacterial blight (caused by *Clavibacter michiganensis*), has caused concern in the United States. The disease thrives under warm conditions. It commonly produces leaf stripe symptoms, but early infection can wither plants (Malvick et al. 2010).

Downy Mildew Diseases of Maize (Peronosclerospora philippinensis)

There are five diseases of maize and sorghum caused by downy mildews:

- *Peronosclerospora philippinensis* (Philippine downy mildew.
- *Peronosclerospora maydis* (Java downy mildew).
- *Peronosclerospora sorghi* (Sorghum downy mildew).
- *Peronosclerospora sacchari* (Sugarcane downy mildew).
- *Sclerophthora macrospora* (crazy top).

Peronosclerospora philippinensis (Fig. 3.8a) is considered to be the most virulent of the downy mildew pathogens affecting maize, causing substantial crop losses ranging from 40% to 100%.

All the above maize downy mildew pathogens show broadly similar symptoms, although there may be variations in intensity depending on the host cultivar, its stage of development and the environment. For example, *P. philippinensis* produces streaks, whereas *P. sacchari* produces blotches on leaves (Murray 2009; dela Cueva et al. 2020).

Young plants, from 2 weeks of age, could show chlorosis and may be stunted. The characteristic mottling, chlorotic streaking and white stripes caused by the disease appear on the leaves of older plants (Fig. 3.8a). Leaves of infected plants are often narrower and more erect in comparison to those of healthy plants and are covered with a white, downy growth on both surfaces.

Crazy top downy mildew of corn is characterized by a proliferation of the tassel (male flowers, Fig. 3.8b), which gives the disease its highly descriptive name. Tassel proliferation, which may be partial or complete, results from the 'replacement of normal floral parts by small leaves: this continues until the tassel resembles a mass of leafy structures' (CPC data sheet, 2005).

Root Rot and Leaf Blight (Pathogen Rhizoctonia solani)

Leaf blight produces 'batik-like' symptoms on the leaves of maize, reminiscent of those on rice.

Common Smut (Pathogen *Ustilago maydis*)

Fleshy smut galls (Fig. 3.8c) of both fungal and plant tissue are formed on leaves, stalks, ears, or tassels of infected plants. Except for those on leaves, these galls contain masses of powdery dark spores, which are released when they mature, dry and rupture (CPC Datasheet, 2005). Control measures include sanitation, seed treatment and agronomic measures. Host resistance is usually the most practical control measure in smut-prevalent areas.

3.2.3 Storypad 4. Not All Fungal Pathogens Are Bad News

The young galls, known as 'Huitlacohe', are a local delicacy in Mexico (and Latin America), where they have been used as human food since Aztec times (León-Ramírez et al. 2014). No immune corn varieties have been found, but field levels of infection are typically low due to the selection of significant levels of host resistance, especially in hybrid corn varieties (Pataky and Snetselaar 2006). Losses can, however, be considerable under conditions of environmental stresses (due perhaps to the meristematic activity following damage) and in crops (e.g. sweet corn, fresh corn) where low levels of contamination with fungal material cause problems with consumer acceptance.

Most introductory courses in plant pathology will offer corn smut because of the easily recognized galls of fungal and host tissue that it produces. It is also an example of the biotrophic stage of a relationship, where host cells are not killed but stimulated by a set of 'effector proteins' (STP proteins) to undergo hypertrophy and hyperplasia, supplying nutrients to the fungus.

Perhaps more importantly, the fungus has been used successfully as a model organism in the studies of the physiology and genetics of host-pathogen interactions since the 1960s, e.g. by exploiting the characteristics of specific mutants (Pataky and Snetselaar, 2006). Its physiology and genetics have also attracted interest as a model in medical studies of tumour development.

The recent sequencing of the genome of *U. maydis* (Brefort et al. 2009) has enabled more progress to be made in proteomics and metabolomics. Much has already been learned about the different mechanisms of disrupting host processes leading to tumour formation, and the set of STP proteins is now known to be unique to *U. maydis*.

From a population genetics perspective, a common question asked is why the fungus does not typically cause more damage when genes for high levels of resistance are not available. Clearly, the fungal genes responsible for producing STP proteins must be subject to intense selection pressures for attributes that increase virulence but have not succeeded in improving epidemic competence. Suggested reasons include superfluous amounts of pollen production and the genetic penalty of diverting resources to gall formation (Ferris and Walbot 2020).

Southern Corn Leaf Blight (Pathogen Cochliobolus heterotrophus)
Symptoms caused by *C. heterotrophus* are usually limited to tan spots on leaves (Fig. 3.8d). During the epidemic of 1970–71 in the United States, race T emerged, causing the coalescing of spots on leaves and extension of symptoms to stalks, leaf sheaths, ear husks, shanks, ears and cobs. A black, felty mold covered affected areas, resulting in severe losses during harvesting and shelling (CPC Datasheet 2005; Bruns 2017).

The most effective method of controlling the disease is to plant resistant hybrids. Cultural control methods include ploughing crop debris into the soil after harvest and rotating crops.

3.2.4 Maize Disease Pictures

Fig. 3.8 Common diseases of maize. (**a**) Downy mildew caused by *Peronosclerospora philippinensis* is probably the most widespread of all the downy mildew diseases on maize. Note the chlorotic streaking of leaves that is characteristic of the disease. (**b**) Crazy top mildew on maize (caused by *Sclerophthora maydis*). Parts of male flowers ('tassel') are replaced by leaves till it forms an untidy mass of leaves or 'crazy top'. (**c**) Common smut of maize (caused by *Ustilago maydis*). Fleshy smut galls formed by a combination of fungal and plant tissue on virtually all aerial parts of the plant, visible here on the tassel. Except for those on leaves, these galls contain masses of almost black, powdery spores that are released as the galls dry. (**d**) Southern corn leaf blight (caused by *Cochliobolus heterostrophus*). Note the tan-coloured elliptical spots on the leaves. Under conducive conditions, the spots coalesce and result in defoliation

References

Bruns HA (2017) Southern corn leaf blight: a story worth retelling. Agron J 109:1–7

Brefort T, Doehlemann G, Mendoza-Mendoza A, Reissmann S, Djamei A, and Kahmann R (2009). Ustilago maydis as a Pathogen. Annual review of phytopathology, 47:423–445. https://doi.org/10.1146/annurev-phyto-080508-081923

Cartwright RD, Groth DE, Wamishe YA, Greer CA, Calvert LA, Cruz CM, Verdier V, Way MO (2018) Compendium of rice diseases and pests. American Phytopathological Society, St Paul, Minnesota. 121 pp

Chi NM, Thu PQ, Nam HB et al (2020) Management of Phytophthora palmivora disease in citrus reticulata with chemical fungicides. J Gen Plant Pathol 86:494–502. https://doi.org/10.1007/s10327-020-00953-z

Chin KM (1974) Chemical control of *Helminthosporium* leaf spot of rice. Malays Agric J 49(4):437–441

Chin KM (1975) Fungicidal control of the rice blast disease. Malays Agric J 50(2):221–228

Chin KM (1976). Occurrence of Thanatephorus cucumeris, the perfect stage of Rhizoctonia solani, on rice in West Malaysia. Mardi Res Bull 4:99–101

References

Chin KM (1977) Chemical control of sheath blight disease of rice caused by *Thanatephorus cucumeris*. Malays Agric J 51(2):238–243

Chin KM (1986) A simple model of selection for fungicide resistance in plant pathogen populations. Phtopathology 77:666–669

Chin KM (2005) Rice disease datasheets. Crop protection compendium. CAB International, Wallingford

Chin KM, Ajimilah NH (1982) Rice variety mixtures in disease control. In: Proceedings of the international conference of plant protection in the tropics, Kuala Lumpur, pp 241–246

Chin KM, Supaad MA (1986) Diseases of rice in Malaysia. MARDI, Kuala Lumpur, p 89

Chukwu SC, Rafii MY, Ramlee SI et al (2019) Bacterial leaf blight resistance in rice: a review of conventional breeding to molecular approach. Mol Biol Rep 46:1519–1532. https://doi.org/10.1007/s11033-019-04584-2

Cruz CD, Magarey RD, Christie DN, Fowler GA, Fernandes JM, Bockus WW, Valent B, Stack JP (2016) Climate suitability for *Magnaporthe oryzae* Triticum pathotype in the United States. Plant Dis 100:1979–1987

Dela Cueva FM, de Castro AM, de Torres RD (2020) *Peronosclerospora philippinensis* (Philippine downy mildew of maize). Invasive species compendium. CABI, Wallingford. https://doi.org/10.1079/ISC.44646.20210200696

Dyer PS, Kück U (2017) Sex and the imperfect fungi. Microbiol Spectrum 5(3): FUNK-0043-2017. https://doi.org/10.1128/microbiolspec. FUNK-0043-2017

Ferris AC, Walbot V (2020) Understanding Ustilago maydis Infection of Multiple Maize Organs. J Fungi (Basel). 2020 Dec 27;7(1):8. https://doi.org/10.3390/jof7010008. PMID: 33375485; PMCID: PMC7823922

Habibuddin H (1978) Incidence of rice ragged stunt disease of rice in Malaysia. MARDI Research Bulletin 6(1):113–117

IRRI (2007) A hybrid history. Rice Today 6(4). International Rice Research Institute DAPO Box 7777, Metro Manila, Philippines Web: ricetoday.irri.org

IRRI (2016) Rice that changed the world: celebrating 50 years of IR8. Rice Today, International Rice Research Institute DAPO Box 7777, Metro Manila, Philippines Web: ricetoday.irri.org

IRRI (2024) Rice knowledge bank. Diseases. http://www.knowledgebank.irri.org/step-by-step-production/growth/pests-and-diseases/diseases

Kumar R, Srinivas K, Boiroju NK, Gedam PC (2014a) Production performance of maize in India: approaching an inflection point. Int J Agric Statist Sci 10(1):241–248

Kumar R, Srinivas K, Monayem Miah MA, Shah H (2014b) Assessment of the maize situation, outlook and opportunities in Asia. In: Conference: 12th Asian maize conference and expert consultation on maize for food, feed, nutrition and environmental security, Bangkok

León-Ramírez C, Sánchez-Arreguín J and Ruiz-Herrera J (2014). Ustilago maydis, a Delicacy of the Aztec Cuisine and a Model for Research. Natural Resources, 5:256–267. https://doi.org/10.4236/nr.2014.56024

Lim GS (1972) Studies on penyakit merah disease of rice. III Factors contributing to an epidemic in North Krian Malaysia. Malays Agric J 483:278–294

Liu Z, Zhu Y, Shi H, Qiu J, Ding X, Kou Y (2021) Recent progress in rice broad-spectrum disease resistance. Int J Mol Sci 22(21):11658. https://doi.org/10.3390/ijms222111658

Malvick D, Syverson R, Mollov D, Ishimaru CA (2010) Goss's bacterial blight and wilt of corn caused by *Clavibacter michiganensis* subsp. *nebraskensis* occurs in Minnesota. Plant Dis 94(8):1064. https://doi.org/10.1094/PDIS-94-8-1064A

Murray GM (2009) Threat specific contingency plan: Philippine downy mildew of maize (*Peronosclerospora philippinensis*) and downy mildew of sorghum (*P. sorghi*). Australia: Plant Health Australia.https://www.planthealthaustralia.com.au/wp-content/uploads/2013/03/Downy-mildew-of-maize-and-sorghum-CP-2009.pdf

Ning X, Yunyu W, Aihong L (2020) Strategy for use of rice blast resistance genes in rice molecular breeding. Rice Sci 27(4):263–277

Ou SH (1985) Rice diseases, 2nd edn. Commonwealth Mycological Institute, Kew, p 380

Pataky J, Snetselaar K (2006) Understanding Ustilago maydis Infection of Multiple Maize Organs. The Plant Health Instructor 6. https://doi.org/10.1094/PHI-I-2006-0927-01

Puckridge DW, Kupkanchanul T, Palaklang W, Kupkanchanaku K (2000) Production of rice and associated crops in deeply flooded areas of the Chao Phraya delta. In: Proceedings of a conference on deep-water Rice and associated crops, Bangkok

Renfro BL, Ullstrup AJ (1976) A comparison of maize diseases in temperate and in tropical environments. PANS 22(4):491–498

Song J-H, Wang Y-F, Yin W-X, Huang J-B, Luo C-X (2021) Effect of chemical seed treatment on rice false smut control in field. Plant Dis 105:3218–3233. https://doi.org/10.1094/PDIS-11-19-2411-RE

Yuan S, Linquist BA et al (2021) Sustainable intensification for a larger global rice bowl. Nat Commun 12:7163. https://doi.org/10.1038/s41467-021-27424-z

Zhu Y, Chen H, Fan J, Wang Y, Li Y, Chen J, Fan J, Yang S, Hu L, Leung H, Mew TW, Teng PS et al (2000) Genetic diversity and disease control in rice. Nature 406(6797):718–722. https://doi.org/10.1038/35021046

Chapter 4
Field Crops

Abstract Field crops are not always consistently defined in the literature. For the purposes of this chapter, I shall therefore define field crops as field-grown, *annual* crops, which include FAO (2017) defined crop groupings like (dry) legumes, oilseeds, root and tuber, spice and beverages and 'others', but exclude cereals and vegetables which are covered in other chapters.

The discussion here includes tobacco, oilseeds (soya beans and groundnuts), legumes (mung beans) and root crops (cassava). Mention is made of other field crops in the appendices.

See Appendices for disease control measures.

4.1 Tobacco

4.1.1 Production

In 2014, the five leading tobacco-growing countries were China, Brazil, India, the United States and Indonesia. China alone produced nearly half of the global total of 7.5 million tonnes (FAOSTAT 2014).

The three major tobacco types are Virginia, burley and oriental. These tobaccos are grown in over 30 countries, including Argentina, Brazil, China, Greece, Italy, Malawi, Mozambique, Spain, Tanzania, Turkey and the United States.

Virginia, or flue-cured tobacco, is also known as 'bright tobacco' because of the golden-yellow to deep-orange colour it takes on during curing. Typically cured for a week in heated barns, it has a light, bright aroma and taste. Virginia tobacco is mainly grown in Argentina, Brazil, China, India, Tanzania and the United States.

Burley tobacco is light to dark brown in colour. Air-cured in barns for up to 2 months, burley loses most of its natural sugars and develops a strong, almost cigar-like taste. It is mainly grown in Argentina, Brazil, Italy, Malawi and the United States.

Oriental tobacco is highly aromatic. Its small leaves are harvested individually and sun-cured in the open air. It is mainly grown in Bulgaria, Greece, Macedonia and Turkey.

4.1.2 Itinerary

Mexico

Tobacco has been grown in Southern Mexico since the 1500s, producing three cultivars: Black Jaltepac, Black San Andrés and Sumatra. Most of this production is intended for export to Europe, the United States and Canada to produce high-quality cigars. Increasingly, however, local brands based on combining the best fillers suited for their quality San Andrés wrappers have gained popularity in global markets (Fig. 4.1a).

4.1.3 Storypad 5: The 'Black Gold' of the San Andrés Valley, Mexico

This area of rich volcanic soils (Fig. 4.1b) on the Gulf of Mexico produces some of the finest dark-coloured, aromatic tobacco for cigars, locally known as black gold. The leaves are greatly in demand for binders as well as wrappers for Maduro-style cigars in the Sierra de los Tuxtlas region of Mexico's Gulf Coast, as well as for other hand-rolled cigars worldwide.

The crop is at a threat from two fungal diseases, blue mould and black shank, as well as several viral diseases. Because of the potential damage to leaves, management of blue mould disease caused by Peronospora tabacina during the wet months of November to January, is critical for the production of high-quality tobacco.

Tobacco fields tend to be mono-cropped and continuously planted without a break, conditions which favour the build-up of blue mould populations.

In some parts of the world (e.g. some farms in Cuba), the plants are shaded with muslin to induce production of the larger leaves that are favoured for cigar manufacture. The reduced light and increased humidity, however, favour blue mould incidence, thus requiring potential intervention measures with fungicides.

4.1.4 Tobacco Disease Panorama

Blue Mould of Tobacco (Caused by Peronospora hyoscyami f. sp. tabacina)
Blue mould is probably the most important leaf disease of tobacco (Borras-Hidalgo et al. 2010). Symptoms of the disease usually begin with the occurrence of yellow

spots (Fig. 4.2b) on the upper leaf surface, which coalesce into dark necrotic lesions resulting in the deformation of the leaves or defoliation. A typical blue-grey downy layer of sporulation (Fig. 4.2c) develops on the lower leaf surface. For quality reasons, leaves damaged by blue mould can no longer be used as cigar wrappers.

A systemic infection of the stem can cause stunting of the plants.

Severe losses from epidemics have been reported both in Central America and the United States. In the warm-temperate/Mediterranean areas of Europe, it occurs regularly each year. CORRESTA (Cooperation Centre for Scientific Research Relative to Tobacco) organizes the Euro-Mediterranean Blue Mould Information Service to collect and share information on the spread of the disease each season.

Black Shank of Tobacco (Caused by Phytophthora nicotianae)
The initial symptom of the disease is a sudden wilting of the leaves, which turn yellow and dry up. This is followed by blackening of the roots and base of the stem (Fig. 4.2a). In the nurseries, the disease can occur as damping-off of the plantlets.

Infection by chlamydospores or zoospores from infected soil requires a temperature above 20 °C. In the field, infection usually comes from transplanting infected seedlings. Attention to nursery and field sanitation is, therefore, key.

4.2 Groundnuts

Extended reading: Kokalis-Burelle et al. (1997), Culbreath et al. (2002), Shew (2020), Hartman et al. (2016).

4.2.1 Production

Groundnuts (peanuts), *Arachis hypogaea* are native to South America. They have had a long history of cultivation, probably predating the present Amazonia by several millennia (Coffelt and Simpson 1997). The leading producers are China and India, although the United States has the highest yield per hectare (Table 4.1).

Table 4.1 Leading producers of groundnuts in 2020

Country	Groundnuts production in 2020 (mio tonnes)
China	14.4
India	7.3
Nigeria	3.2
USA	2.1
Sudan	2.0
Indonesia	1.2

Groundnuts are reputedly one of the most nutrient-dense foods, with about 25% protein and 50% lipids, most of which are poly- or mono-unsaturated fatty acids (IFT 2019).

Aflatoxin exposure due to *Aspergillus* infection/contamination can, however, be important under conditions of subsistence farming in the tropics, where storage conditions are suboptimal (see Aflatoxins in Groundnuts).

The threat from aflatoxins in peanuts (and other major food crops like maize and tree nuts) contaminated with *Aspergillus flavus* and *A. parasiticus* is well known. In the United States, strict enforcement of storage conditions has largely removed the threat, but aflatoxin continues to be a challenging topic under conditions of subsistence farming in the tropics, where these crops are staples. IPFRI (Unnevehr and Grace 2013) has estimated that 26,000 people in Sub-Saharan Africa die annually from liver cancer associated with aflatoxin exposure. Other health effects of exposure include immunity suppression and child stunting. Aflatoxins also affect international markets and trade (e.g. loss of market access due to product contamination).

4.2.2 A Tale from the Sub-Sahara by Scheuring, JF (pers comm. 2022)

Aflatoxin exposure is particularly problematic in low-income populations. A. flavus infection is especially acute in dry conditions towards the end of the season, particularly on poor, drought-prone soils in the Sub-Sahara. A common sight in the street markets of many small Sub-Saharan towns is the grading of groundnuts for sale, with the large, plump nuts priced much higher than those which are smaller and darker in colour...yet another sad example of 'the poor getting poorer and sicker'.

4.2.3 Storypad 6: Ancient Survival Packages Still Relevant Today

Groundnuts (A. hypogaea) produce aerial flowers, but then unlike most other plants, form their pods and seeds well protected below ground. About a week after self-fertilization, a 'peg' develops and pushes the ovary into the ground. The tip then moves into a horizontal position before the fruit is produced.

Other survival features include a deep rooting system and an ability to pause their growth at times of drought, thus enabling the plant to conserve water. Its ability to fix nitrogen reduces dependence on added nitrogen as well as the overall carbon footprint of the crop.

This ancient survival kit ticks the box on many of our present agronomic needs. Interestingly, it was one of the crops that has helped the regeneration of soils in the SE United States that had been depleted from intensive cotton cultivation during the late nineteenth to early twentieth century.

Perhaps better known for enjoying refuge in the soil is the root crop cassava (Manihot esculenta Crantz), which can withstand drought, fire, frost, poor soil nutrition, feeding by wild animals and the rapine that often follows unrest and war (Otang et al. 2023). The survival strategy of the crop is based on its fleshy tuberous roots, which enable it to be stored in the ground for almost a year. The high yield produced per unit ground area (an astonishing 50 to 82 tonnes per ha, according to Jansz and Uluwaduge 1997), compared to other carbohydrate-producing crops, has made it a major source of food in Africa, Latin America and Asia the crop where it is key to food security.

4.3 Soybeans

4.3.1 Production

Genomic studies have recently confirmed that the earliest domestication of the crop occurred in China (Sedivy et al. 2017). Soybeans (*Glycine max*) are today the world's most important legume crop, used both as a high-protein source of food and feed, as well as in a large number of industrial and consumer applications. Fueled by feed demands from increasing meat consumption in developing economies and the widespread popularity of vegan diets, global demand has soared. Like all other legumes, soybeans can fix their own nitrogen, accounting for an average of 55% of their needs (Wortmann et al. 2019), when nodulation is not hindered (e.g. by inadequate amounts of other nutrients like phosphorus and potassium). The amount of nitrogen fixed is inversely related to its availability in the soil (Rahmat et al. 2023), so that the amount and timing of adding fertilizers must be carefully calibrated with crop needs at different stages of growth.

The following table tells the story of the rapid ascendency of Brazil as the world's leading producer in 2022. Overtaking production in the United States in 2019–20, the surge was all the more impressive because of the high cost of controlling Asian Soybean Rust (ASR) since 2001–02 (Table 4.2).

Table 4.2 Leading producers of soybeans in 2012 and 2022

Country	mio Ha		mio tonnes	
	2012	2022	2012	2022
Brazil	25.0	40.9	65.8	120.7
USA	30.8	34.9	82.8	116.4
Argentina	17.6	15.9	40.1	43.9

4.4 Mung Beans

4.4.1 Production

Mung beans (*Vigna radiata*) are native to the Indian subcontinent and widely cultivated in South Asia. The leading producers are India and Myanmar.

4.4.2 Storypad 7: Not Just for Beansprouts

Mung beans are generally considered in most countries to be a minor crop but have gained global interest recently because of the increasing popularity of plant-based proteins and vegan egg products. Grown in 4–6 days as sprouts, it is a quick source of fresh vegetables. Perhaps more importantly, the crop is tolerant to drought and high temperatures.

*These features enable the crop to meet the two important environmental challenges of our time, sustainability and climate change. Food benefits include easy digestibility (hypoallergenic and free of flatulence), as well as high nutritional and potential pharmacological value (*Batzer et al. 2022*).*

4.5 Cassava

4.5.1 Production

Cassava (*Manihot esculenta*) is native to South America. The leading countries producing cassava are Nigeria, the Democratic Republic of Congo, Thailand, Ghana, Brazil and Indonesia (FAOSTAT 2023).

Global yields per hectare of cassava in 2023 were on average less than that of potatoes but exceeded those of maize, rice or wheat. These high yields and the resilience of the crop under adverse conditions (see above, *Storypad 6*) have made cassava a key crop for food security in many countries in the tropics (Janz and Uluwaduge 1997). It is also a major source of animal feed and industrial products. The use of cuttings to propagate new plants reduces the cost of replanting but suffers from the disadvantage of resulting in homogeneous stands of the crop (see virus diseases below).

4.5.2 Legume and Root Crop Disease Panorama

Passalora (Formerly Cercospora) Leaf Spot of Groundnuts
Extended reading: Culbreath et al. (2002), Woodward et al. (2013), Shew (2020).

4.5 Cassava

4.5.3 Groundnuts diseases

Early leaf spot and late leaf spot are the most common and important peanut diseases in North Carolina, United States. The diseases cause defoliation and yield loss of 50% or more. *Nothopassalora personata* (*Cercosporidium personatum*) causes late leaf spot, and *Passalora arachidicola* (*Cercospora arachidicola*) causes early leaf spot.

Early leaf spots are typically brown (Fig. 4.3a), whereas late leaf spots are dark brown in colour, most easily visible from the underside surface of a leaf. Lesions are initially round but become irregular in shape when they coalesce. Early leaf spot lesions are generally round but can expand or grow together, becoming less regular in shape. They can be found as early as 30 days after planting.

Late leaf spot symptoms are first seen on leaves at the bottom of the plant. In severe cases, the plants can be completely defoliated and die. Stems may also bear lesions (Shew 2020).

4.5.4 Soybeans diseases

Asian Soybean Rust (ASR)

Soybean and its diseases have been extensively reviewed by various authors (e.g. Hartman et al. 2016; Godoy et al. 2016). Some stories are, however, worth re-telling because they are a replay of other familiar diseases caused by the rust fungi (see *Storypad 1*). For example, the three major rust diseases of wheat (leaf, stem and stripe/yellow) cause damage worldwide and are a major threat to food security (Lidwell-Durnin and Lapthorn 2020). In the same way, coffee rust can devastate entire plantations of coffee. Spread by wind over large distances, these rusts produce explosive epidemics capable of causing total crop loss. They are often associated with the classic 'boom and bust cycles' that are well known in plant pathology. Cultivars with new resistance genes boom in popularity and then experience the bust when virulent strains capable of overcoming the host resistance are selected for and become dominant, setting the scene for the next boom cultivar.

The ASR disease (caused by *Phakopsora pachyrhizi*) was first detected in Brazil in 2001 (Godoy et al. 2016). It was first reported in the United States in 2005, and in the short period of 1 year had spread to eight southern states of the country. The speed of defoliation is reminiscent of rust diseases in other crops. However, unlike rusts of other crops, where survival between crops in winter can be restricted by the few species of alternative hosts, a wide range of plants serve this purpose for ASBR (Lidwell-Durnin and Lapthorn 2020).

A wide range of control tactics have been tested in Brazil, but the major tool for the present is still based on fungicides applied preventively. In the United States, application is timed with the first sign of incoming spores or based on infection on trap plants exposed a few weeks before the commercial crop. As in the case of yellow rusts of wheat and powdery mildew of barley, the genetic plasticity of ASBR, means that breeders working with resistance genes must be vigilant about which ones to deploy each season. To counter this threat, Johnson (1983) introduced the concept of durable resistance in wheat to yellow rust.

Interestingly, *Rhizoctonia solani* (*T. cucumeris*), which causes rice sheath blight disease, also attacks soybeans. Fields that rotate rice with soybeans, therefore, tend to have a greater incidence of the disease.

4.5.5 Mung Bean diseases

Cercospora Leafspots

Leafspots caused by *Cercospora* spp. are first round, water-soaked and brown (Fig. 4.3b). These coalesce later into irregular, gray lesions with narrow reddish margins. The centre of a lesion becomes detached, giving a shot-hole effect. Under warm, wet conditions, extensive defoliation occurs. *Cercospora* also attacks other tropical legumes like cowpeas and soybeans (Bughio 2017).

4.5.6 Cassavadiseases

Viral Diseases of Cassava

Like other crops (e.g. bananas) which are vegetatively reproduced, cassava stands tend to be genetically homogeneous. We have discussed in previous chapters how this homogeneity encourages the dissemination and spread of diseases and pests. Viral diseases of cassava are often the major threat to the crop. Cassava mosaic disease in Africa and Asia, brown streak disease in Africa and frog skin disease in Latin America are transmitted through the use of infected planting material and by vectors (e.g. whiteflies, *Bremia tabaci*). The vector is largely managed through the use of insecticides, although progress is being made to develop entomophagous fungi for its biological control (Sani et al. 2020).

Whilst phytosanitary measures and PCR-based diagnostics have improved disease management, recent advances in genomics offer hope for the development of host resistance to viral diseases (Legg et al. 2015; Jacobson et al. 2018; Otang et al. 2023). Predictive models of disease epidemiology have also been responsible for improving disease management in Africa (Godding et al. 2023).

4.5 Cassava

4.5.7 Field Crop Disease Pictures

Fig. 4.1 Tobacco production in Mexico. (**a**) Visit to tobacco farm and cigar factory, Tuxtla, Mexico: workers seen rolling cigars using 'black gold' tobacco grown in the San Andrés Valley. (**b**) Tobacco field on rich, dark volcanic soil of the San Andrés Valley, Gulf of Mexico

Fig. 4.2 Tobacco diseases. (**a**) Black shank of tobacco caused by *P. nicotianae* attacking the base of the stem. (**b**) Chlorotic spots, later turning brown, of blue mould (caused by *Peronospora*) on the upper leaf surface. (**c**) Bluish sporulating lesions of *Peronospora* on the lower side of the leaf

Fig. 4.3 Diseases of field crops caused by *Cercospora*. (**a**) Peanut leaf spot (caused by *Passalora arachidicola*, formerly *Cercospora arachidicola*). (**b**) *Cercospora* leaf spot of mung beans

References

Batzer JC, Singh A, Rairdin A, Chiteri K, Mueller DS (2022) Mungbean: a preview of disease management challenges for an alternative U.S. cash crop. J Integr Pest Manag 13(1):4. https://doi.org/10.1093/jipm/pmab044

Borras-Hidalgo O, Thomma BPHJ, Yussuan S, Chacon O, Pujol M (2010) Tobacco blue mould disease caused by *Peronospora hyoscyami* f. sp. *tabacina*. Mol Plant Pathol 11(1):13–18

Bughio MA (2017) Cercospora leaf spot in mung beans. CABI International, Plantwise

Coffelt TA, Simpson CE (1997) Origin of the peanut. In: Kokalis-Burelle N, Porter DM, Rodriguez-Kabana R, Smith DH, Subrahmanyam P (eds) Compendium of peanut diseases. APS Press, St. Paul

Culbreath AK, Brenneman TB, Kemerait RC Jr (2002) Management of early leaf spot of peanut with pyraclostrobin as affected by rate and spray interval. Plant Health Progress 3. https://doi.org/10.1094/PHP-2002-1018-01-RS

Godding D, Stutt ROJH, Alicai T, Abidrabo P et al. (2023) Developing a predictive model for an emerging epidemic on cassava in sub-Saharan Africa. http://www.nature.com/scientificreports

Godoy CV, Seixas CDS, Soares RM et al (2016) Asian soybean rust in Brazil: past, present, and future. Pesq Agrop Brasileira 51(5):407–421. https://doi.org/10.1590/S0100-204X2016000500002

Hartman GL, Rupe JC, Sikora EJ, Domier LL, Davies JA, Steffey KL (2016) Compendium of soya bean diseases and pests. American Phytopathological Society, St Paul

Institute of Food Technologists (2019) The most important nut is a legume. Food Technology Magazine 73(10)

Jacobson L, Duffy S, Sseuragi P (2018) Whitefly transmitted viruses threatening cassava production in Africa. Curr Opin Virol 33:167–176

Jansz ER, Uluwaduge DI (1997) Biochemical aspects of cassava (Manihot esculenta) with special emphasis on cyanogenic glucosides. J Natn Sci Council, Sro Lanka 25(1):1–24

Johnson R (1983) Genetic background of durable resistance. In: Lamberti F, Waller JM, Van der Graaff NA (eds) Durable resistance in crops. NATO advanced science institutes series, vol 55. Springer, Boston. https://doi.org/10.1007/978-1-4615-9305-8_2

Kokalis-Burelle N, Porter DM, Rodríguez-Kábana R, Smith DH, Subrahmanyam P (1997) Compendium of peanut diseases. American Phytopathological Society, USA. ISBN 978-0-89054-218-7

Legg P, Lava Kuma P, Makeshkumar T, Trpathi L, Fergusaon M, Kanju E, Ntawuruhunga CW (2015) Cassava virus diseases. Adv Virus Res 19:85–142

Lidwell-Durnin J, Lapthorn A (2020) The threat to global food security from wheat rust: Ethical and historical issues in fighting crop diseases and preserving genetic diversity. Global Food Security. 26. 100446. https://doi.org/10.1016/j.gfs.2020.100446

Otang NV, Tripathi JN, Kariuki SM, Tripathi L (2023) Cassava molecular genetics and genomics for enhanced resistance to diseases and pests. Mol Plant Pathol 25(1)

Rahmat Z, Sohail MN, Perrine-Walker F, Kaiser BN (2023) Balancing nitrate acquisition strategies in symbiotic legumes. Planta 258, 12. https://doi.org/10.1007/s00425-023-04175-3

Sani I, Ismail SI, Abdullah S, Jalinas J, Jamian S, Saad N (2020) A review of the biology and control of whitefly, Bemisia tabaci (Hemiptera: Aleyrodidae), with special reference to biological control using Entomopathogenic fungi. Insects 11(9):619. https://doi.org/10.3390/insects11090619

Sedivy EJ, Wu F, Hanzawa Y (2017) Soybean domestication: the origin, genetic architecture and molecular bases. New Phytol 214(2):539–553. https://doi.org/10.1111/nph.14418

Shew B (2020) NC state extension publications. Peanut leaf spot. https://content.ces.ncsu.edu/early-leaf-spot-of-peanut-1

Unnevehr L, Grace D (2013) Aflatoxins: finding solutions for improved food safety Focus 20. Brief 1 Nov 2013, IPFRI CGIAR, Research program on Agriculture for Nutrition and Health

Wortmann CS, Kaizzi KC, Maman N, Cyamweshi A, Dicko M, Garba M, ... & Serme I (2019) Diagnosis of crop secondary and micro-nutrient deficiencies in sub-Saharan Africa. Nutr Cycl Agroecosyst 113:127–140. https://doi.org/10.1007/s10705-018-09968-7

Woodward JE, Brenneman TB, Kemerait RC Jr (2013) Chemical control of peanut diseases: targeting leaves, stems, roots, and pods with foliar-applied fungicides. In: Fungicides–showcases of integrated plant disease management from around the world, vol 15, pp 55–76

Chapter 5
Plantation Crops

Abstract Palm trees have been described as the big game of the plant kingdom (Bailey in 1933 quoted by Tomlinson 1961). These trees and other perennial tree crops like rubber and cocoa are similarly impressive, largely because of the total productivity that their longevity delivers. The high value of each individual tree over its productive lifetime, and the high costs of replanting, means that the trees must be diligently nurtured and protected. Bananas are a perennial herb rather than a tree, and new plantings become productive in a relatively short period of time, but optimal production is also dependent on the density of tree populations.
See Appendices for disease control measures. Extended reading: Ploetz (2003).

5.1 Bananas

Extended reading: Ploetz (2005, 2006), De Langhe et al. (2009), Ganry et al. (2012).

5.1.1 Production

The centre of diversity of *Musa acuminata*, from which our modern Cavendish bananas arose, is probably located within the rainforests of Malaysia (Simmonds 1966). It is reputed to be among the earliest fruits to be cultivated by man. The first records of the fruit were apparently those made during Alexander the Great's conquest of India in 327 BC, but the fruit only really took off globally with the advent of better communications and transportation systems in the late nineteenth century.

Bananas are today the world's leading fresh fruit, produced mainly in plantations in Asia, Africa and central and northern parts of South America. The largest global producers are India, China, Indonesia and Brazil, but most of their production is locally consumed. The largest exporters are located in Latin America (Ecuador, Costa Rica, Guatemala, Colombia), and in Asia, the leading export country is the Philippines (FAOSTAT 2020). Assuming a 3 °C increase in global temperatures by

2070, the FAO also estimates that there will be a 50% increase in the area suitable for growing bananas.

Costa Rica, well known today as a hub for ecotourism, is home to three household names for producers of tropical fruits: Chiquita, Del Monte and Dole.

Corbana, the national banana corporation of Costa Rica, is perhaps a unique feature of the local banana industry because it is partly owned and managed by the banana growers themselves. By coordinating R&D to build farming skills, helping to open market access for value-added products, and growing communities, whilst protecting the rich natural environment, Corbana has helped empower all banana growers in Costa Rica in an exemplary, sustainable manner.

5.1.2 Fancy Bananas

Countries that produce for their own consumption are not restricted to cultivars that transport well and therefore grow a much wider spectrum of banana varieties and tastes.

It is common to spot seeded banana trees growing wild in the forests of Malaysia, which is one of the centres of origin for the crop. The country has at least 10 popular varieties (Straits Times 15.04.2014), of which only one, Pisang Mas, has been exported. These varieties show a wide range of colours, tastes, shapes and sizes, suitable for various culinary purposes. The germplasm collection of the Malaysian Agricultural Research and Development Institute has about 200 accessions (How 2002).

Progress has recently been made in developing resistance to both Sigatoka as well as the Panama disease, with wild diploid bananas, mutagenesis of Cavendish cultivars and the aid of genomic studies, but it will require significant investments in an industry still largely based on Cavendish cultivars.

5.1.3 Itinerary: Philippines (Fig. 5.1a)

The Philippines is the eighth largest producer in the world but the largest exporter from Asia. Like other major producers, the main cultivar grown is 'Cavendish'.

A typical view of plantations often includes the ubiquitous cables on which the heavy bunches are transported after harvest (Fig. 5.1b). Population control of bearing stems is critical for achieving desired yields and quality (e.g. long, well-developed fingers). Whilst the standard configuration of one bearing/mother stem and two followers (Fig. 5.1c) 2 months apart, is often used to achieve year-round production, this can be further manipulated to achieve specific production goals.

5.1.4 Storypad 8: Two Tales of How a Fungus Determined the Banana We Eat

The first tale, which has been widely retold (see Ploetz 2005 for a detailed and intriguing account), is about the Gros Michel cultivar of banana, which was popular until the 1950–1960s because it met all the post-harvest quality requirements of an industry dependent on transporting a perishable product on long sea journeys. Gros Michel is, however, highly susceptible to Panama disease caused by Fusarium oxysporum f.sp. cubense (see below). Commercial bananas have to be seedless and are therefore propagated only by vegetative cloning. As a result, the narrow genetic base in the large plantations growing Gros Michel was largely defenseless against the rapid spread of Panama disease.

The cultivar Cavendish is reputed to have been first selected by William Paxton in the greenhouses of Chatsworth House, England, in 1836. He named the selection Musa cavendishii after his employer, the 7th Duke of Devonshire, William Cavendish. 'Cavendish' and its sub-selections were found to possess similar desirable fruit qualities like 'green fruit transportation', and crucially, were resistant to Panama disease. It rapidly replaced Gros Michel in the 1960s.

Race TR4 (Tropical Race 4) of Panama disease is, however, able to attack Cavendish but had previously only been limited to SE Asia. In recent years, it has begun to spread and has been detected, e.g. in Australia, Peru, Colombia and Africa, putting the global industry again on alert for alternative solutions (TR4 Portal, CABI Invasive Species Compendium 2022).

In contrast to the concentration of international trade on single 'Cavendish type' cultivars, bananas grown and consumed locally in Asian Pacific countries are based on a wide variety of cultivars (Ploetz 2006), even though they belong mainly to the two species, Musa acuminata (A genome) and M. balbisiana (B genome). Cavendish is a triploid AAA. The region (including Malaysia, Indonesia, Thailand and the Philippines) accounts for 61% of global consumption (Voora et al. 2020), mainly of locally grown bananas. Freed of the physical constraints of export requirements, consumer preferences are for smaller, sweeter and more aromatic bananas, in a variety of shapes, colours and textures appropriate for different culinary and dessert purposes. In the Philippines, which is the largest exporter of Cavendish from Asia, the favoured local cultivars are Lacatan (AAA) and Lantundan (AAA). In Malaysia, Pisang Raja (AAB), Mas (AA), Rastali (AAB), Berangan (AAA), Abu Bujal (ABB) and Awak (ABB) are some of the most popular (Plotz 2006; Mostert et al. 2017). In India (the largest producer in the world), half the production is from local cultivars (Mostert et al. 2017).

Consumption of locally grown bananas is similarly high in Africa (e.g. Uganda, Rwanda and Cameroon, Vorra et al. 2020). This high genetic diversity of edible bananas for local consumption in the Asia Pacific, their cultivation in relatively small, non-contiguous areas, and evidence of useful levels of host resistance to Fusarium (and sigatoka disease) among some cvs (Plotez 2015; Mostert et al. 2017) may be part of the reason why epidemics of Fusarium have taken longer to take hold, and sigatoka is infrequently treated with fungicides in areas that are outside plantations.

Host resistance or tolerance remains the best hope of controlling Fusarium. Conventional breeding for resistance has been difficult in the past because of polypoidy, genetic abnormalities, sterility and difficulties in maintaining desired aspects of quality and parthenocarpy/sterility (Ortiz and Swennen 2014). PCR-based profiling of regional strains of Fusarium has increased the precision of determining the range of pathogen genotypes that breeders and quarantine pathologists need to manage (Mostert et al. 2017).

5.1.5 Banana Disease Panorama

Sigatoka and Cordana

Major leaf diseases include the Sigatoka diseases (caused by *Mycosphaerella musicola* and *fijiensis*) and *Cordana* leaf spot. There is a new name, *Pseudocercospora*, for *Mycosphaerella*, but I shall use the old name here because it is more easily recognizable.

Sigatoka diseases are widespread in most countries where bananas are grown. Black sigatoka or black leaf streak (BLS caused by *M. fijiensis*, Fig. 5.2) is probably the best known because of the striking symptoms of streaks coalescing to form blackened leaves (Fig. 5.2b). Severely infected fields have been likened by some growers to 'black death' because of the macabre appearance of trees, especially when viewed in inclement weather. *M. musicola* causes yellow sigatoka, exhibiting eyespot lesions and yellow-brown dried tissue (Fig. 5,3b). Yellow sigatoka is usually less aggressive than BLS, but may be more widespread in some countries, like Brazil (Brito et al. 2015).

Sigatoka infection is spread by conidia (short distance by splash dispersal, Fig. 5.2c) and by ascospores (Fig. 5.2a), long distance by wind. The disease badly affects crop yields both by reducing the development of fruit bunches, as well as affecting the keeping quality of fruits from infected trees (e.g. premature ripening). Workers are therefore trained to monitor trees that may require fungicidal treatment, and at harvest to count the number of 'functional leaves', rejecting fruit from trees that have, e.g. four or fewer functional leaves. These trees yield less and are also likely to have quality issues that are detrimental to fruit export.

Whilst cultural methods and fungicides have been effectively used for the control of black sigatoka in plantations for the last few decades, selection for resistant races able to overcome specific modes of action is a continuing challenge. To counter this challenge, resistance management strategies of fungicides, not unlike those for managing pathogen resistance in other crops, have been developed (Chin et al. 2001). Ganry et al. (2012) have also stressed the importance of integrated control.

The eyespot lesions of another common leaf disease (caused by *Cordana*) are much larger, not unlike the mythological eyes of 'Cyclops' (Fig. 5.3c). It is, however, less destructive than sigatoka.

5.1 Bananas

5.1.6 Storypad 9. Aleyandro's Tale: A Day in the Life of a Banana Plantation Manager in Costa Rica

Aleyandro (fictitious name) is the manager of a large banana plantation in Costa Rica. To maintain the high productivity of his crop, he has a team of scouts whose job it is to 'keep a finger on the pulse of his trees', especially during the rainy months from June to the end of January. Aleyandro is, however, a college graduate and likes to be personally involved in farm decision-making. He, therefore, takes time to check the statistics they provide him and ensure that mitigation measures are planned and executed when the numbers deviate from acceptable ranges.

He understands that crop nutrition, plant density, water management and disease control are among the key tools he needs to optimize yields and maintain the required fruit quality for export. On his morning walks through selected parts of his plantation, he therefore habitually checks on each of these areas, especially during the rainy months from the end of May to the end of January when most of the 4000 mm of annual rain falls. On the agronomic side, he looks for signs of nutrient deficiency/imbalance and water stress. To ensure that he has the optimal tree density to meet his production targets, his field force is trained to keep the number of followers of each mother plant to 2 per plant and ensure that most trees have a minimum of four functional leaves (i.e. leaves with less than 30% necrosis) per plant. He is also conscious of the importance of disease prevention rather than cure. Literature (Marin et al. 2003) tells him that necrotic lesions of leaves on plants can produce spores for up to 22 weeks. As a sanitary measure, his workers have been routinely removing dead leaves from the trees.

Closer to flowering, he counts the number of leaves that are free of streaks from the top youngest leaf to know if the fungicide applications he ordered have been effective in controlling BLS. His target is to have at least four youngest leaves free of black leaf streaks. 'Youngest Leaf Streaked' could be just 2 when control has been poor. Wherever possible, he practices integrated control measures to reduce the cost of production (e.g. see Ganry et al. 2012). Later in the day, Aleyandro reviews his observations with his team and plans the work schedule for the following week. It is getting late in the day, but he has one more important call to make. His last stop is his processing and packaging plant, where he checks if the current batch of harvested fruit has had their 'bath' of disinfectant and been examined to be free of crown rot and other post-harvest diseases before their long journey to Europe.

5.1.7 Storypad 10. Size Helps. A Whole Replicate of a Field Trial Can Fit on a Single Leaf

Banana leaves are widely used in the tropics as waterproof plates to serve food and as aromatic wrappers for cooking. Another use for which the size of the leaves has been put to good use is in reducing the cost of field experimentation. They are so large that an entire replication of a fungicidal trial against sigatoka could be conducted by spray treatment of 'micro-plots' marked out on a single leaf.

Panama Disease

An infected plant may appear healthy with no discolouration for 2 months to a year, and then the petiole collapses, leaves turn yellow and the whole plant wilts and dies (Fig. 5.3a). The blocked vascular bundles, stained red or purple, are visible when the pseudo-stem is cut and emits a fishy odour. The fungal pathogen *Fusarium oxysporum* is soil-borne, and entry is often facilitated by nematode damage. Control by sanitation or chemicals has not been successful. The plant tries to defend itself by producing multiple suckers, but these are invariably infected as well (PANS Manual No. 1 1971).

Banana plants can also lodge following wind damage (Fig. 5.3d), but this damage is distinguishable by the badly torn leaves and absence of wilting.

Crown Rot Disease (Post-harvest)

Quality concerns are reflected in the strict post-harvest processes that the fruit is subjected to prepare them for their journey to the market. For example, fruits are disinfected in water baths and checked for physical damage and post-harvest diseases like crown rot and anthracnose, before packing for export. The storage conditions on the journey are carefully monitored. A similar set of observations and checks to those before departure are repeated at the port of entry for comparative and quality purposes.

5.2 Oil Palm

Extended reading: Chung (2011).

The main cultivated species is *Elaeis guineensis* (African oil palm). *Elaeis oleifera* (American oil palm) is also grown for its higher oleic acid content. LH Bailey famously called palms the 'big game of the plant kingdom' (quoted by Tomlinson 1961). Unusually for a monocotyledon plant, they are woody and certainly among the largest in that group of plants. The tallest palms can reach 60 m in height. Oil palms can reach 20 m in height but are usually replaced before they get to a third of that height to facilitate easy harvesting of the fruits (Jones and Hughes 1989). Fruit bunches can weigh up to 30 kg.

5.2.1 Production

The leading countries growing oil palm in 2020 were Indonesia, Malaysia and Thailand. Oil is extracted from both the pericarp and the kernels of the fruit. Colombia is the largest grower in Latin America. Worldwide, in 2020/2021, 72 million metric tonnes of palm oil were produced, making up 31.4% of the world's oil and fat production (FAOSTAT).

5.2 Oil Palm

Table 5.1 Leading producers of oil palm in 2020 (FAOSTAT)

Country	Palm oil bunches production in 2020 (mio tonnes)
Indonesia	256.5
Malaysia	97.0
Thailand	15.7
Nigeria	9.5
Colombia	7.2
Guatemala	2.9

Based on average data, the oil yield per hectare of oil palm (4.1 tonnes per ha) can be 8-nine-fold that produced by cotton, peanut, soybean and rapeseed (Murphy 2014; Johnston et al. 2009). Because the crop is perennial, production costs per hectare of crop are also considerably less than those from annual crops. The understory is usually managed with legume cover crops to protect soil structure (Table 5.1).

5.2.2 Oil Palm, Rubber and Cocoa Itinerary

Plantations in East and West Malaysia.

5.2.3 Oil Palm Disease Panorama

Basal Stem Rot

Basal stem rot, caused by *Ganoderma boninense*, is the most important disease of oil palm in SE Asia (Paterson 2006a). Other diseases predominate in Africa (*Fusarium* wilt) and in Latin America (bud rot).

Initially, wilting and signs of malnutrition occur. Young fronds may remain unopened at the spear stage, allowing light to penetrate the canopy (Fig. 5.4b). Telltale green vegetation sprouts on the ground surface around an infected tree. Subsequently, older fronds die back, and sporophores (mushrooms) are formed at the base (Fig. 5.4c, d).

Palms may be killed (Fig. 5.4a) within a year of symptoms appearing on the leaves (Kranz et al. 1977).

5.2.4 Storypad 11. Pathogen and Panacea

Ganoderma boninense may be a dreaded and costly pathogen in oil palm plantations, but cultivated mushrooms (Lingzhi) of a related species, Ganoderma lucidum, have been used for medicinal purposes in China for over 2000 years (Li et al. 2019).

In 1995, the total estimated annual market value of cultivated Ganoderma, given by different commercial sources, was US$1628 million (Chang and Boswell 1999).

It is today one of the most prescribed ingredients in Traditional Chinese Medicines in the far east. DNA analyses have recently been employed to understand the phylogenetic relationships of the different species of the genus Ganoderma (Zhou et al. 2015). Paterson (2006b) has reviewed the range of metabolites produced that have shown potential therapeutic benefits.

5.2.5 Storypad 12. Shining a Light with LAMP: Model Specifications for Field Diagnostic Technology

We have already encountered Lethal Yellowing disease of another key palm tree, coconuts, in Chap. 1. The prompt removal and destruction of infected palms before the pathogen can spread is the only practical way of control. The corresponding rapid field diagnosis needed to back up sanitation was, however, problematic, because the disease is symptomless in the incubation phase, and Phytoplasmas are virtually impossible to grow in axenic culture.

Shining a lamp on the problem, Dickinson's team at Nottingham University, United Kingdom, developed LAMP (Loop-mediated isothermal amplification, Dickinson 2015) based on nucleic acid detection. Its specifications read like what should be a model of requirements for all field technologies. It is quick (2 min DNA extraction, 20 min for the whole test, faster than usual PCR tests), robust (minimum equipment, battery-operated, relatively impure DNA needed, reagents do not require refrigeration) and precise (guards against false negatives, and primers have been developed for country-specific lethal yellowing pathogens).

Fusarium oxysporum f. sp. *elaeidis (FOE), causing the wilt disease of oil palm, is another 'cryptic' pathogen that has been successfully diagnosed in the field using LAMP technology (Adusei-Fosu and Dickinson 2023). 'Lamp' assays have greatly improved the diagnosis of the Cape St. Paul wilt disease (a lethal yellowing-like disease of coconut).*

LAMP is used at present not only for diagnostics of viral, fungal and bacterial pathogens but also for food testing against bacterial and fungal contaminants (Niesson 2015).

5.3 Rubber

(Extended reading: Rao 1975)

The rubber tree (*Hevea braziliensis*) is native to the Amazonian region of South America. Its commercial cultivation in local plantations failed despite several attempts, due in part to the South American Leaf blight disease. Imported material into Malaysia (which, due to strict phytosanitary measures, remains free of the disease till today) enabled its rapid adoption of the crop, and success following high demand for natural rubber for automobile tires from the early 1900s. Critical to the technological development of natural rubber were the establishment of industrial standards pioneered by the Rubber Research Institute Malaysia.

Tapping of the rubber tree for latex is highly labour-intensive. In recent years, the plantation sector in Malaysia has shifted more to oil palm because of higher profitability and lower labour requirements. Other countries like Thailand, Indonesia and Vietnam now lead the world in the production of natural rubber.

5.3.1 Production (Table 5.2)

Itinerary: Malaysia (Fig. 5.5a).

5.3.2 Rubber Tree Disease Panorama

As the flow of latex is related to plant health, diseases of rubber have a direct impact on its commercial productivity. Perhaps the most common diseases in Asia are the root diseases and *Phytophthora palmivora*, which causes abnormal leaf fall, dieback of new shoots and black stripe of the tapping panel (Drenth and Guest 2013; Krishnan et al. 2019; Misman et al. 2022).

White, Red and Brown Root Rots
In Malaysia, diseases affecting the root system have been the most destructive. These include *Rigidoporus lignosus*, *Ganoderma pseudoferreum* and *Phellinus noxius*, causing respectively the white, red and brown root diseases due to the colour of their 'rhizomorphs'/aggregations of fungal tissue on the tree roots (Rao 1975).

White root (Fig. 5.5b–e) is probably the most common. Symptoms include discolouration of the leaves followed by defoliation. Branches then die back, and the trees ultimately die.

Table 5.2 Leading producers of natural rubber in 2020 (FAOSTAT)

Country	Rubber (natural) production in 2020 (mio tonnes)
Thailand I	4.7
Indonesia	3.4
Vietnam	1.2
India	1.0
Côte d'Ivoire	0.9
China	0.7

Unearthing the roots will reveal the fungal 'rhizomorphs', which appear as white flattened strands on the infected roots (Fig. 5.5b, c). Fructifications occur as orange-coloured fleshy brackets (Fig. 5.5e) at the collar region of the stems or exposed roots. Treating the collar region of a young, infected tree with fungicidal paint (Fig. 5.5d) may appear laborious, but it is justified by the 30-year productive lifespan of an average tree.

5.4 Cocoa

Cocoa (*Theobroma cacao*) is native to Central America, where it has been cultivated since the time of the Mayans and Aztecs as a valuable food (broma) of the gods (theo). It was imported by the Spanish conquistadores into Europe as a mild stimulant but did not gain widespread popularity until ways were found to improve its palatability.

Cocoa is a typical understory inhabitant of forests, with the survival of young trees requiring the protection of shade and humidity from taller surrounding trees. As a result, new plantations are intercropped with other cash/food crops like coconut and bananas for at least 3 years. The income from these plantings, as well as other activities like animal husbandry, helps tide small-holder farmers over the period before the cocoa becomes productive.

5.4.1 Production (Table 5.3)

Itinerary: Sabah Malaysia (Fig. 5.6a).

5.4.2 Cocoa Disease Panorama

The most widespread disease is black pod disease (Fig. 5.6b), caused by *Phytophthora palmivora*. Other diseases are more location-specific, e.g. witches broom and frosty pod rot occur more in Latin America; swollen shoot in Africa; and vascular streak dieback and pink disease in SE Asia.

Table 5.3 Leading producers of cocoa, 2020 (FAOSTAT)

Country	Cocoa beans production in 2020 (mio tonnes)
Côte d'Ivoire	2.2
Ghana	0.8
Indonesia	0.7
Nigeria	0.3
Ecuador	0.3
Cameroon	0.3

5.4 Cocoa

Pink disease, caused by *Corticium salmonicolor*, has a wide host range, attacking economic crops (e.g. rubber, coffee, tea, citrus, durian) as well as ornamentals like *Hibiscus* and a number of legumes (Rao 1975). Its pink incrustations are formed from mycelia and basidia on infected branches, which die back as the disease progresses.

Vascular Streak Dieback (VSD) Disease
VSD is caused by a fungal pathogen, *Ceratobasidium theobromae* (syn. *Oncobasidium theobromae*) (Guest and Keane 2018).

5.4.3 Storypad 13. The Cryptic Pathogen of Cocoa

The cryptic symptoms of VSD consist of leaf yellowing (with green islands) and shedding, followed by dieback of shoots and ultimate death of nursery seedlings and trees. They occur with no apparent sign of any fungal growth, bacterial exudates or nematode damage (Fig. 5.7a, b). As a result, researchers were misled for years into thinking that the disease was viral or nutritional in origin.

The hidden pathogen was finally determined by Keane in 1981. Only when leaves drop off, and in wet weather, does the fungus reveal itself by producing basidiocarps (fruiting bodies) at the leaf scars (Fig. 5.7d). Xylem vessels are also visibly stained when an infected stem is split (Fig. 5.7c), and the stem surface is rough due to raised lenticels.

At a late stage of disease development, trees are stunted and die back.

Black Pod Disease
Black pod of cocoa is caused by several *Phytophthora* species, but *palmivora* is perhaps the most common part of the species complex. All parts of the plant may be attacked, but the most damaging are rotting of cocoa pods and discolouration and shriveling of the cocoa beans (Bowers et al. 2001). The lesions start as dark brown to black spots, which spread to cover the whole fruit (Fig. 5.6b). The fruits become mummified and serve as a source of infection for several years. Typically, it causes 20% to 30% pod losses through black pod rot and kills up to 10% of trees annually through stem cankers (Fig. 5.6c) (Guest 2007).

P. palmivora is a ubiquitous pathogen in the tropics and attacks several other crops like coconut, durian and papaya.

5.4.4 Plantation Crop Disease Pictures (Figs. 5.1, 5.2, 5.3, 5.4, 5.5, 5.6, and 5.7)

Fig. 5.1 Growing bananas. (**a**) Visit to banana plantation, Philippines: note the bagging of fruit bunches to hasten maturity and protect from pest damage. (**b**) Cable transportation of harvested banana bunches to the processing plant. (**c**) The three generations of banana. Production in a plantation is largely managed by controlling the population density of trees. Effectively, this is done by restricting the number of followers. A popular configuration is a mother plant followed by two followers/suckers, spaced about two months apart

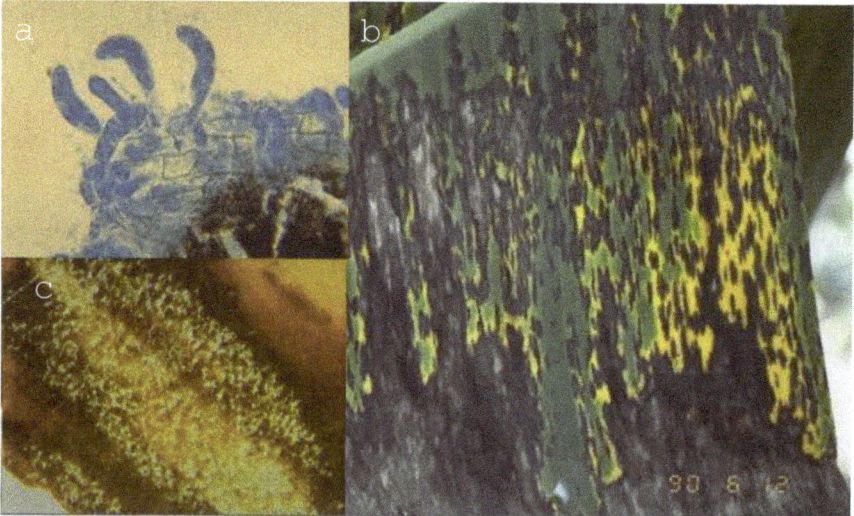

Fig. 5.2 Black sigatoka, black leaf streak. (**a**) Ascospores (sexual state) of *Mycosphaerella* (syn. *Pseudocercospora) fijiensis*, the causal organism of black sigatoka or black leaf streak disease of banana. These spores enable the long-range, wind dispersal of the pathogen under dry weather conditions. (**b**) Black sigatoka disease lesions on banana leaf, usually more obvious on the lower surface of the leaf. Characterized initially by the linear streaks of black lesions aligned with the leaf veins; these eventually coalesce to form the black, dried areas on each leaf. Symptoms in severely affected plantations may be so ominous they have been described as looking like the 'black death' (PANS 1971). (**c**) Conidia of *M. fijiensis*: Responsible for the short-range splash dispersal of the pathogen in wet weather

5.4 Cocoa

Fig. 5.3 Other common maladies of banana. (**a**) Panama disease (pathogen *Fusarium oxysporum* f. sp. *cubense* [Foc]) causing wilting of banana trees. (**b**) Yellow sigatoka or yellow leaf spot caused by *Mycosphaerella musicola*. Spots are brown to yellow-brown in colour, narrower and more obvious on the upper leaf surface (*cf. black leaf streak*). Damage is typically less than that caused by black sigatoka. (**c**) Leaf spots caused by *Cordana musae*. Typically, eye-shaped, brown lesions, becoming light grey in colour, surrounded by a darker margin and concentric zones. (**d**) Lodging of banana trees following a typhoon in Taiwan

Fig. 5.4 Basal stem rot disease of oil palm. (**a**) Tree killed by basal stem rot disease, caused by *Ganoderma boninense*. (**b**) The first indication of basal stem rot disease in an oil palm plantation is a gap in the canopy due to the stunting of leaf fronds. At ground level (not shown), the appearance of green patches of vegetation around the base of the tree is another indication of a thinning canopy. (**c**) Mushrooms of *Ganoderma boninense* appear around the base of the infected tree. (**d**) Basidiocarp (mushroom) of *G. boninense*. A related species, *G. lucidum (lingzhi in Chinese; reishi in Japanese)*, is widely reported to have multiple health benefits

Fig. 5.5 Diseases of rubber. (**a**) Young rubber trees in a plantation. (**b**) Rubber seedlings attacked by white root disease, caused by *Rigidoporus lignosus*. (**c**) Fresh infection of 'white root' unearthed in the field. (**d**) Treating the collar region of an infected tree with fungicidal paint. (**e**) Older infection developing orange basidiocarps

Fig. 5.6 Phytophthora disease of cocoa. (**a**) Visit to Cocoa Plantation, Sabah, Malaysia. (**b**) Black pod of cocoa caused by *Phytophthora palmivora*. (**c**) *Phytophthora* lesions on the stem revealed when the bark is removed

Fig. 5.7 Vascular streak dieback (VSD) disease of cocoa. (**a**) discolouration of leaves (appearing like a nutrient deficiency) and dieback of shoots, but no external sign of fungal infection. (**b**) dieback of shoots allows sunlight to penetrate the leaf canopy. A green patch of grass in a cocoa plantation often marks the location of a tree suffering from VSD. (**c**) split stem shows stained vascular tissue. (**d**) *the fungal pathogen Oncobasidium theobromae* finally reveals itself, emerging from leaf scars

References

Adusei-Fosu K, Dickinson M (2023) Detecting Fusarium oxysporum f. sp. elaeidis by using loop-mediated isothermal amplification. J Plant Pathol 105:1637–1643. https://doi.org/10.1007/s42161-023-01435-9

Bowers JH, Bailey BA, Hebbar PK, Sanogo S, Lumsden RE (2001) The impact of plant diseases on world chocolate production. Plant Health Progress 2(1):12. https://doi.org/10.1094/PHP-2001-0709-01-RV

Brito F, Fraaije B, Robert M (2015) Sigatoka disease complex of Banana in Brazil: management practices and future directions. Outlooks Pest Manag 26(2):78–81

CABI (2022). TR4 Portal. Invasive Species Compendium. Wallingford, UK: CAB International. www.cabi.org/isc

Chang S-T, Boswell J (1999) *Ganoderma lucidum* (Curt.: Fr.) P. Karst. (Aphyllophoromycetideae)–a mushrooming medicinal mushroom. Int J Med Mush 1:139–146

Chin KM, Wirz M, Laird D (2001) Sensitivity of *Mycosphaerella fijiensis* from Banana to trifloxystrobin. Plant Dis 85:1264–1270

Chung G (2011) Management of Ganoderma diseases in oil palm plantations. Planter 87:325–339

De Langhe E, Vrydaghs L, de Maret P, Perrier X, Denham T (2009) Why bananas matter: an introduction to the history of banana domestication. Ethnobot Res Appl 7:165–177

Dickinson M (2015) Loop-mediated isothermal amplification (LAMP) for detection of Phytoplasmas in the field. In: Lacomme C (ed) Plant pathology. Methods in molecular biology, vol 1302. Humana Press, New York, p 99. https://doi.org/10.1007/978-1-4939-2620-6_8

Drenth A, Guest DI (2013) Phytophthora palmivora in tropical tree crops. In: Lamour K (ed) Phytophthora: a global perspective. Centre for Agriculture and Bioscience International (CABI), pp 187–196. https://doi.org/10.1079/9781780640938.0187

Ganry J, Fouré E, de Lapeyre de Bellairaire L, Lescot T (2012) An integrated approach to control the black leaf streak disease (BLSD) of bananas, while reducing fungicide use and environmental impact. In: Fungicides for plant and animal diseases, pp 194–226. https://doi.org/10.5772/29794

Guest D (2007) Black pod: diverse pathogens with a global impact on cocoa yield. Phytopathology 97:1650–1653

Guest D, Keane P (2018) Cocoa diseases: vascular-streak dieback. In: Achieving sustainable cultivation of cocoa, pp 287–301. https://doi.org/10.19103/AS.2017.0021.18

How TC (2002) Classification of banana varieties in Malaysia using retrotransposons. MSc thesis. Universiti Putra Malaysia

Johnston M, Foley JA, Holloway T, Kucharik CJ, Monfreda C (2009) Resetting global expectations from biofuels. Environ Res Lett 4:014004. https://doi.org/10.1088/1748-9326/4/1/014004

Jones LH, Hughes WH (1989) Oil Palm (Elaees guineensisJacq.). In: Biotechnology in Agriculture and Forestry, Vol.5, Trees II. Y.P.S. Bajaj (ed.), Springer-Verlag, Heidelberg, pp. 176–202

Kranz J, Schmutterer H, Koch W (1977) Diseases pests and weeds in tropical crops. Verlag Paul Parey, Berlin, p 666

Krishnan A et al (2019) An insight into Hevea – Phytophthora interaction: the story of Hevea defense and Phytophthora counter defense mediated through molecular signalling. Curr Plant Biol 17:33–41. https://doi.org/10.1016/j.cpb.2018.11.009

Li Z, Zhou JL, Lin Z (2019) Development and innovation of Ganoderma industry and products in China. Adv Exp Med Biol 1181:187–204

Marín DH, Romero RA, Guzmán M, Sutton TB (2003) Black Sigatoka an increasing threat to banana cultivation. Plant Dis 87(3):208–222

Misman N, Samsulrizal NH, Noh AL, Wahab MA, Ahmad K, Ahmad Azmi NS (2022) Host range and control strategies of *Phytophthora palmivora* in Southeast Asia perennial crops. Pertanika J Trop Agric Sci 45(4):991–1019

Mostert D, Molina AB, Daniells J, FourieG, Hermanto C, Chao C-P, et al. (2017) The distribution and host range of the banana Fusariumwilt fungus, Fusarium oxysporum f. sp. cubense, inAsia. PLoS ONE 12(7): e0181630. https://doi.org/10.1371/journal.pone.0181630

Murphy DJ (2014) The future of oil palm as a major global crop: opportunities and challenges. J Oil Palm Res 26(1):1–24

Niessen L (2015) Current state and future perspectives of loop-mediated isothermal amplification (LAMP)-based diagnosis of filamentous fungi and yeasts. Appl Microbiol Biotechnol 99(2):553–574. https://doi.org/10.1007/s00253-014-6196-3

Ortiz R, Swennen R (2014) From crossbreeding to biotechnology-facilitated improvement of banana and plantain. Biotechnology advances, 32(1):158–169. https://doi.org/10.1016/j.biotechadv.2013.09.010

PANS Manual No.1 (1971) Pest control in bananas. Overseas Development Administration, London, p 128

Patterson RRM (2006a) Ganoderma boninense (basal stem rot of oil palm). Cabi Compendium, Datasheet. https://doi.org/10.1079/cabicompendium.24924

Patterson RRM (2006b) Ganoderma – a therapeutic fungal biofactory. Phytochemistry 67(18):1985–2001. https://doi.org/10.1016/j.phytochem.2006.07.004

Ploetz RC (ed) (2003) Diseases of tropical fruit crops. Cabi Publishing, p 481

Ploetz RC (2005) Panama disease, an old nemesis rears its ugly head: part 1, the beginnings of the banana export trade. Plant Health Progress 6. https://doi.org/10.1094/PHP-2005-1221-01-RV

Ploetz RC (2006) Panama disease, an old nemesis rears its ugly head: part 2, the cavendish era and beyond. Plant Health Progress. https://doi.org/10.1094/PHP-2006-0308-01-RV

Rao BS (1975) Maladies of *Hevea* in Malaysia. Rubber Research Institute of Malaysia, p 108

References

Simmonds NW (1966) Bananas, 2nd edn. Longmans, London, p 512

Tomlinson PB (1961) Anatomy of the Monoctyledons Vol. 2: Palmae. Edited by C. R. Metcalfe. Pp. xiii + 453 + 9 plates. (Oxford: Clarendon Press; London: Oxford University Press, 1961.)

Voora V, Larrea, C and Bermudez S. (2020). Global market report: Bananas (p. 12). International Institute for Sustainable Development (IISD)

Zhou XW, Su KQ, Zhang YM (2015) Phylogenetic analysis of widely cultivated *Ganoderma* in China based on the mitochondrial V4-V6 region of SSU rDNA. Genet Mol Res 14(1):886–897

Chapter 6
Vegetables

Abstract Vegetables are perhaps the most universal and diverse group of crops we will encounter on our journey. Even though one subgroup, the legumes, are virtually independent of added nitrogen, most vegetables are highly demanding of high levels of soil fertility. The diversity of vegetable species and the high fertilizer inputs typical of vegetable culture have attracted an equally wide range of plant pathogens.

Production
See Appendices for disease control measures.
Extended reading: Koike et al. (2003), Ebert (2017), Muniappan and Heinrichs (2016), Srinivasan (2016) (Table 6.1).

6.1 Solanaceous Vegetables and Beans

Extended reading: Jones et al. (2016).

Tomatoes, peppers and aubergines together made up about 24% of the world's total production of fresh vegetables in 2020, with tomatoes alone accounting for 16% (FAOSTAT). Tomatoes and peppers are of New World origin, whereas aubergines originated from the Old World (Asia and Africa).

Table 6.1 Leading producers of fresh vegetables, 2020 (FAOSTAT)

Country	Vegetables fresh production in 2020 (mio tonnes)
China	170.4
India	40.2
Vietnam	15.4
Nigeria	6.8
Philppines	5.2
Nepal	4.0

Table 6.2 Lead producers of tomatoes in 2020 (FAOSTAT)

Country	Tomato production in 2020 (mio tonnes)
China	64.9
India	20.6
Turkey	13.2
USA	12.2
Egypt	6.7
Italy	6.2

6.1.1 Production (see Table 6.2)

6.1.2 Solanaceous Vegetables and Beans Disease Panorama

Diseases of solanaceous vegetables tend to be soil-borne in origin, perhaps because of the lack of crop rotation under conditions of protected culture. As a result, they are the most common vegetable crops to be commercially grafted, using disease- and nematode-resistant rootstocks to protect the scion.

Grafting can also increase vigour and productivity by increasing the number of plants per rootstock and improving salt tolerance. Such practices are likely to decrease in importance as the areas under soil-less cultivation (e.g. hydroponics) increase, and advances in genomics produce plants with improved disease resistance.

Anthracnose and Cercospora Diseases
We shall encounter anthracnose (caused by *Colletotrichum gloeosporiodes*) repeatedly in our journey because it attacks a wide range of crops in the tropics. It is a common disease on pepper, where it can cause severe fruit damage (Fig. 6.1b). We have also encountered *Cercospora* before, on rice and mung beans. On pepper, *Cercospora* causes leaf spots (Fig. 6.1c), leading to defoliation in severe cases.

Phytophthora Diseases
Phytophthora spp. are the most common pathogens on Solanaceous vegetables. *Phytophthora capsici* causes foot and fruit rot and crown blight of pepper (Fig. 6.1d). Infected plants wilt and ultimately die.

6.1 Solanaceous Vegetables and Beans

Late blight of tomato is caused by *Phytophthora infestans* (Fig. 6.3b; which we have encountered earlier on potatoes in *Storypad 1*). It causes a similarly devastating blight.

Early Blight and Other Diseases

The early blight pathogen, *Alternaria solani*, also produces similar symptoms (spots with concentric rings commonly called *'bullseye'*) on both tomato (Fig. 6.3d) and potato leaves.

Other important diseases include *Rhizoctonia* blight (we have encountered the pathogen before on rice and maize) on both tomato (Fig. 6.3e) and aubergines (Fig. 6.4a); and tomato powdery mildew (Fig. 6.3c) (*Oideum neolycopersici*) and bacterial canker of tomato (Fig. 6.3a) (*Clavibacter michiganensis*).

Diseases Caused by Nematodes

Rootknot (*Meloidogyne* spp.) is the most common nematode occurring on all Solanaceous vegetables, including tomatoes, peppers and aubergines. Other frequent hosts include okra and some leafy vegetables. Major species of rootknot include *M. incognita, M. javanica, M. arenaria and M. hapla* (Seid et al. 2015). Apart from the obvious root knots (Fig. 6.4d), infected plants are stunted and often wilted. Symptoms are, however, not always clear on the harvested part of a crop, e.g. potato tubers may be symptomless (Moens et al. 2009).

Host resistance, rotation with non-host crops, sanitation and biocontrol are often the only practical solutions besides the use of fumigants and chemicals.

Some success has been achieved against *M. arenaria, M. incognita* and *M. javanica* with breeding for host resistance using the *Mi-1* gene in tomatoes. M. *enterolobii* (*guava nematode*) is, however, an emerging threat capable of infecting tomato cultivars with the *Mi-1* resistance gene (Seid et al. 2015).

6.1.3 Storypad 14: A Tiny Pest of Superlative Proportions

We have not mentioned plant-parasitic nematodes before in this book, but this often microscopic-sized soil pest can be full of superlatives. According to van den Hoogen et al. (2019), *'approximately 0.3 gigatonnes of nematodes inhabit surface soils across the world', making it 'the most abundant form of animal life on this planet'. Rootknot nematodes have also been reported to parasitize nearly all species of higher plants (*Moens et al. 2009*).*

*Rootknot is one of the most important diseases of tomato, the world's leading vegetable crop. It is found wherever the crop is grown but is especially important in warm temperate to tropical regions. One might almost add 'however' the crop is grown because it is a major problem, even in the shallow sandy soils of the well-known Almeria region of Spain, where vegetables are produced under protected 'greenhouse-like' conditions (*Fig. 6.4d*), as well as in hydroponic cultures.*

6.2 Cucurbit Vegetables

Extended reading: Keinrath et al. (2017).

Cucumbers and gherkins accounted for about 8% of the world's fresh vegetable production in 2020 (FAOStat) (Table 6.3).

6.2.1 Cucurbit Disease Panorama

Downy Mildew

Downy mildew (causal agent, *Pseudoperonospora cubensis*) is probably the most common disease of cucurbits in the tropics. It is a wet-weather disease and causes rapid defoliation of the plant. The characteristic symptoms are yellow, angular leaf spots (demarcated by the leaf veins) on the upper leaf surface (Fig. 6.2b) and corresponding water-soaked lesions on the lower surface. The lesions coalesce, and plants have a burnt look. Grey downy mildew sporulations appear on the lower leaf surface.

Gummy Stem Blight

Gummy stem blight (*Didymella bryoniae*) attacks seedlings, leaves, the stem and fruits of cucurbits. Typically, water-soaked lesions form on the stem, ultimately girdling it (Fig. 6.2c). A reddish-brown gummy liquid exudes from the cankers on the stem.

Cucumber Foot Rot

Foot rot (Fig. 6.2a). is caused by *Fusarium oxysporum*, which we have visited before on bananas. Typical of the damage caused by the pathogen, infected plants wilt and rapidly die. Control is challenging, although soil disinfection and grafting onto resistant rootstocks have been found to be effective (Pavlou et al., 2002).

Table 6.3 Leading producers of cucurbits and gherkins, 2020 (FAOStat)

Country	Cucumber and Gherkin production in 2020 (mio tonnes)
China	72.8
Turkey	1.9
Russian Federation	1.7
Iran (Islamic Republic)	1.2
Mexico	1.2
Ukraine	1.0

6.2.2 Other Vegetable Diseases Panorama

Extended reading: Schwartz and Mohan (2016).

Vegetables are too diverse a group to attempt to cover in a satisfactory manner here. Aside from solanaceous vegetables and cucurbits, it is only intended to mention other crop diseases frequently observed in visits to vegetable farms in the tropics.

These include purple blotch (another *Alternaria*, this time *A. porri*) on shallots (Fig. 6.4b) and onions, leaf spot/blight of cabbage (*A. brassicicola*) with its typical bullseye spots (Fig. 6.4c) and powdery mildew (*Erysiphe cichoracearum*) of sesame. As is perhaps evident in the picture (Fig. 6.4e), powdery mildew of sesame can be a particularly destructive disease, causing up to 50% yield loss.

6.2.3 Vegetable Disease Pictures

Fig. 6.1 Diseases of pepper. (**a**): healthy pods of pepper. (**b**): pepper infected with anthracnose (caused by *Colletotrichum gloeosporiodes*). (**c**): Cercospora leaf spots on pepper. (**d**): *Phytophthora capsici* causing foot rot and crown blight on pepper

Fig. 6.2 Diseases of cucurbits. (**a**): cucumber foot rot caused by *Fusarium oxysporum*. (**b**): downy mildew of cucumber caused by *Pseudocercospotrella cubensis*. (**c**): *Didymella* gummy stem blight of cucurbits

Fig. 6.3 Diseases of solanaceous vegetables. (**a**): bacterial canker of tomato. (**b**): *Phytophthora* spores from late blight of tomato. (**c**): powdery mildew of tomato. (**d**): early blight of tomato. (**e**): *Rhizoctonia* blight of tomato

Fig. 6.4 Other fungal and nematode diseases of vegetables. (**a**): *Rhizoctonia* blight of aubergines. (**b**): purple blotch of shallot (caused by *Alternaria porri*); also attacks onions. (**c**): *Alternaria* blight of cabbage (typical bullseye spot). (**d**): rootknot nematode disease on tomato (right). Healthy roots on the left. (**e**): powdery mildew (*Erysiphe cichoracearum*) of sesame

References

Ebert AW (2017) Vegetable production, diseases, and climate change. In: World agricultural resources and food security: international food security. Emerald Publishing Limited, pp 103–124

Jones JB, Zitter TA, Momol TM, Miller SA (2016) Compendium of tomato diseases and pests. American Phytopathological Society, St Paul

Keinrath AP, Wintermantel WM, Zitter TA (2017) Compendium of cucurbit diseases and pests. American Phytopathological Society, St Paul, Minnesota

Koike S, Subbarao K, Davis RM, Turini T (2003) Vegetable diseases caused by soilborne pathogens. UCANR Publications

Moens M, Perry RN, Starr JL (2009) Meloidogyne species: a diverse group of novel and important plant parasites. In: Perry RN, Moens M, Starr JL (eds) Root-knot nematodes. CABI Publishing, Wallingford, pp 1–17

Muniappan R, Heinrichs EA (eds) (2016) Integrated pest management of tropical vegetable crops. Springer

Pavlou GC, Vakalounakis DJ, Ligoxigakis EK (2002) Control of root and stem rot ofcucumber, caused by Fusarium oxysporum f. sp. radicis-cucumerinum, by grafting onto resistant rootstocks. Plant Dis 86:379–382

Schwartz HF, Mohan SK (2016) Compendium of onion and garlic diseases and pests. American Phytopathological Society, St Paul

Seid A, Fininsa C, Mekete T, Decraemer W, Wesemael WML (2015) Tomato (Solanum lycopersicum) and root-knot nematodes (Meloidogyne spp.)– a century-old battle. Nematology 17:995–1009

Srinivasan R (2016) Integrated pest management in tropical vegetable crops. In: Integrated pest management in the tropics, pp 219–247

van den Hoogen J et al (2019) Soil nematode abundance and functional group composition at a global scale. Nature 572(7768):194–198. https://doi.org/10.1038/s41586-019-1418-6

Chapter 7
Fruit Crops

Abstract The volume of tropical fruits in world trade has been growing rapidly in recent years, largely because of improvements in transportation and shifts in consumer preferences. Aside from bananas, the vast majority (95%, FAO, Major tropical fruits—statistical compendium 2021. Rome, 2022) of the four major tropical fruits, mangoes (including mangosteens and guava), avocados, pineapples and papayas, are however still locally consumed.
See Appendices for disease control measures.

Extended reading: Ploetz (2003).

7.1 Mango

All cultivated mangoes belong to the single species *Mangifera indica*. The centre of origin is in the Indo-Thai-Malaysia-Myanmar region (Mukherjee 1972). It is widely cultivated in subtropical to tropical areas. India is by far the largest producer, but Mexico is the leading exporter.

Fruit set is triggered in India by the dry season, but in SE Asia the flowering stimulus is less clear (Davenport 2007). Uneven fruit set is in general *the* major problem for commercial production and can be aggravated by disease problems like anthracnose, which attacks the inflorescence and causes premature fruit drop.

The genetics of mango is complicated by different degrees of polyembryony (e.g. in India and SE Asia) and apomixis. This has resulted in a large number of cultivars being grown that differ in flavour, shape, size and colour (Burkill 1966).

Table 7.1 Leading producers of mangoes, guavas and mangosteens, 2020 (FAOSTAT)

Country	Mangoes, guavas, mangosteens production in 2020 (mio tonnes)
India	24.7
Indonesia	3.6
China	2.5
Mexico	2.4
Pakistan	2.3
Brazil	2.1

7.1.1 Production (Table 7.1)

7.1.2 Mango Disease Panorama

Anthracnose of Mango

Anthracnose of mango, caused by *Colletotrichum gloeosporioides*, is found in virtually all mango-growing areas of the world. Leaves (Fig. 7.1d), flowers and fruits (Fig. 7.1b, c) of the tree are attacked. Young shoots turn black and die back. Round brown spots appear on leaves resulting in deformation and premature leaf-fall. Inflorescences are desiccated and turn black (Kranz et al. 1977). Sunken, round, black lesions typical of anthracnose (Fig. 7.1c) appear on fruits, causing premature fruit drop, or coalesce and cause rotting of the fruits.

A feature of the pathogen on fruits (and several other hosts) is that it can, after penetration into the host, remain asymptomatic and latent for months and only develop anthracnose upon ripening after harvest. This feature can cause commercial problems since mangoes are usually harvested mature but not ripe.

No commercial cultivars that are resistant to the disease are currently available, and fungicides are the main means of control (Dofuor et al. 2023). Post-harvest treatment of the fruits has also been reported.

Colletotrichum gloeosporioides has a wide range of hosts in tropical crops. For example, it also attacks avocado, banana, beans, citrus, coffee, guava, mangosteens, papaya and wax apple.

7.2 Durian

Durians (*Durio zibethinus*) are native to the rainforests of Malaysia, Indonesia and Thailand. The leading producers are Thailand, Malaysia and Indonesia. Thailand is however by far the largest exporter and China accounts for 95% of global imports (FAO 2023a, b).

7.2.1 Storypad 15. The Legendary Fruit of SE Asia and Its Nemesis?

Durians are arguably the most highly valued fruit in SE Asia. They have become well known today for their many extravagant attributes (extraordinary taste, smell, nutritional and medicinal value). As can be attested by a casual web search, they have also attracted perhaps the most colourful quotations about their qualities. My own favourite descriptions come from the two evolutionary biologists to whom we owe our basic understanding and principles of natural selection.

On a trip to Borneo in 1856 Alfred Russel Wallace called it the king of fruits: 'a rich custard highly flavoured with almonds', 'flavours that call to mind cream-cheese, onion-sauce, sherry-wine', 'a rich glutinous smoothness in the pulp' and 'the more you eat of it the less you feel inclined to stop'. Darwin, on the other hand, is reputed to have said that it is something you 'would wish that your worst enemy is forced to eat'.

To add to the complexity of its attributes, Burkill (1966) notes that the aroma and taste of durian change rapidly each day after the fruit is harvested. Interestingly, he also noted that it is such a highly valued fruit that some indigenous tribes would clear the forest around selected wild trees and camp nearby so that they could be sure of collecting fruits at the perfect stage of ripeness. Nevertheless, all commercial airlines have, for a long time, universally decided to ban its carriage within a plane cabin. Interestingly, many who found durians initially unbearable have been converted to the latter class after repeated exposure.

Folklore has it that durians have such an intensive effect on the senses that they should never be consumed accompanied by alcoholic drinks. Another is that the only way to get rid of the smell that remains on the hands, after a meal of durians, is to wash them with water contained in the shells of the fruit. Durians are also said to be so potent that a spike in some local populations may be noticeable each year, nine months after the fruit season.

The insatiable demand for popular, highly prized cultivars like Musang King of Malaysia and Monthong (Golden Pillow) of Thailand among durian connoisseurs has resulted in a booming export market, mainly to East Asia. In 2021–2022, Thailand exported US$3.3 billion worth of durian fruits, making it the country's third-largest export commodity after natural rubber and rice (FAO 2023a). The strong domestic demand within Malaysia and Indonesia still accounts for most of the local produce, from a wide variety of native cultivars.

It is therefore hardly surprising that disease threats to the crop are taken very seriously by growers, especially when the pathogen, Phytophthora palmivora, is another legendary player in tropical crop protection (see below).

7.2.2 Durian Disease Panorama

Durian Patch Canker and Dieback

Phytophthora palmivora causes a wide range of symptoms on durian. It causes root rot, patch canker of the stem (Fig. 7.2b), dieback of branches (Fig. 7.2a) and fruit rot. A 25% yield loss and high rates of mortality have been reported in orchards (Singh et al. 2025). The most recognizable symptom, patch canker, is characterized by a reddish-brown patch, necrosis on the bark and reddish-brown discolouration of the sapwood (Fig. 7.2b), often with gummy exudates. The entire tree may be killed if the lesion girdles the trunk. Loss of trees can be catastrophic, taking into consideration average yield estimates of 100–200 fruits per tree per year and a 60-year productive lifespan of a healthy tree (FAOStat 2023).

Control in the past has concentrated on cultural methods like ensuring adequate soil drainage, managing soil pH and organic amendments, sanitation, as well as the use of resistant clones, and fungicide application (painting of cankers/trunk injections/soil drenches). Progress in the use of host resistance has been hampered by the lack of a rapid method of progeny testing. Recent work (Noor Camillaet al. 2019), with molecular markers, has offered the hope of greater precision in the screening of progeny for resistance breeding.

We have already encountered the same ubiquitous pathogen on cocoa. Other important hosts of *Phytophthora* among tropical crops include papaya (fruit rot), pineapple (heart rot), citrus (canker), black pepper (foot rot) and coconut (bud rot) (Erwin and Ribeiro 1996).

7.3 Other Fruits

7.3.1 *Pineapple: A Fruit Fit for Kings*

Pineapple (*Ananas comosus*) is native to tropical America and was introduced to the world by the Portuguese and Spaniards after the discovery of the New World. As a fruit fit for kings, it is often depicted in stories and pictures of its first presentation to kings and emperors, e.g. in England, Spain and India (Burkill 1966). The poor keeping quality of the fruit on long sea journeys without the aid of refrigeration meant that only the wealthy could afford to have them (grown in greenhouses) in Europe until the advent of the twentieth century.

The largest producers are currently Indonesia, the Philippines and Costa Rica (FAOStat 2023). Leading exporters are Costa Rica, the Philippines and the Netherlands (which exports most of what it imports).

Heart and foot rots (caused by fungi *Phytophthora cinnamoni*, *Pythium* and by the bacterium *Dickeya zeae* syn. *Erwinia chrysanthemi*) are perhaps the most important of several diseases that attack pineapples (Anderson et al. 2012; Sapak and Nusaibah 2024).

7.3.2 Papaya *(Carica Papaya)*

Papaya is also native to the Americas. India, the Dominican Republic and Indonesia currently lead the world in producing the fruit. Leading exporters of papaya in 2023 were, however, Mexico, Guatemala and Brazil.

A feature of the tree is the production of copious amounts of latex, especially from young tissue. Besides its excellent qualities as a dessert fruit, Burkill (1966) noted several traditional culinary and therapeutic/pharmacological uses, as well as the extraction of the well-known enzyme papain. Perhaps the first feature of papaya that students of botany will learn is that it exemplifies dioecious plants, i.e. male and female flowers are produced on separate plants.

Papaya Ring Spot Virus is probably the most important disease of papaya. It is characterized by ring spotting of fruits, as well as mosaic and chlorosis symptoms on leaves. The virus is transmitted by an aphid. Papaya mosaic virus disease was formerly thought to be a distinct disease but has now been confirmed to be caused by the same virus (Encyclopedia of Virology 2008). Anthracnose is also a common disease.

7.3.3 A Tale About Three Apples That Really, Are Not Apples

Wax apple, rose apple and custard apple are all tropical tree fruits. The taxonomy of these and related fruits, like cloves, is complex, and work is still in progress to clarify relationships. The genus *Syzygium* alone accounts for more than a thousand species (Soh 2017).

Wax apple or Java apple (*Syzygium samarangense*) is a tropical fruit tree plant in the family of Myrtaceae. It is bell-shaped, usually white to pink in colour, and has 'fluffy' flesh with a distinctive aroma of roses. It probably originated in the SE Asian region but is today grown all over the tropics. Formerly known as *Eugenia javanica,* or in Malay very aptly as 'jambu ayer mawar' or rose water jambu (Burkill 1966).

Water or rose apple also belongs to the same genus but is named *Syzygium acqueum* or *malaccense (*Malay rose apple*)*. Formerly known as *Eugenia aquea* (Burkill 1966), it is more oblong-shaped, mild-flavoured and usually red in colour.

Sugar or custard apple (*Annona reticulata*) is native to tropical America but has naturalized in many tropical countries. The thick segmented rind, white flesh and seeds are arranged in a symmetrical layer around a core. A related fruit, soursop (*Annona muricata*), has a slightly acidic taste and is also native to tropical America (Burkill 1966).

Anthracnose of Wax Apple
I have personally encountered anthracnose (*Colletotrichum*) only on wax apple but would be surprised if the disease does not occur on all three 'tropical apples'. The sunken, circular lesions caused are typical of damage caused by the pathogen. We have encountered anthracnose before on banana, pepper, mango, papaya and beans.

7.3.4 Tale of a Hairy Fruit and Its Relatives

Rambutans (*Nephelium lappaceum*), and some of its wild relatives, are aptly described as hairy fruits. The fruits are commonly red in colour, but some cultivars have yellow fruits. Some of the related wild species with long hair are thought to be responsible for the folklore that they are used as braided hair by some specsies of monkeys (Burkill 1966).

A similar and equally popular local fruit, pulasan (formerly *Nephelium mutabile*, now *N. ramboutan-ake*), has the same characteristic white, sweet flesh (aril), but the fruit is covered by short spines instead of hairs. Both rambutan and pulasan are native to the Malay Peninsula. Perhaps better known in international trade are the litchis (lychees) and longans, which belong to the same family (Sapindaceae) and also have the characteristic fleshy aril around each seed.

Fruit rot of rambutan, caused by a complex of fungi in a number of countries, has been described by Zakaria (2022).

7.3.5 Fruit Disease Pictures (Figs. 7.1 and 7.2)

Fig. 7.1 Mango anthracnose disease. (**a**) Visit to mango orchard, Malaysia. (**b**) Anthracnose attacking young fruits, causing fruit drop. (**c**) Typical sunken, round to oval, black lesions, with concentric rings of spores on inoculated mango (bottom fruit). (**d**) Severe anthracnose on leaves leading to defoliation

Fig. 7.2 Durian patch canker disease. (**a**) Tree dieback due to *Phytophthora palmivora* infection in a durian orchard. Poor soil drainage is often a key factor that facilitates infection. (**b**) The disease typically attacks the tree trunk towards the base of the tree, as well as roots, causing a canker with sticky exudates. Removing the canker on the bark will reveal a reddish to brown-coloured stain on the exposed wood.

References

Anderson JM, Pegg KG, Scott C et al (2012) Phosphonate applied as a pre-plant dip controls Phytophthora cinnamomi root and heart rot in susceptible pineapple hybrids. Australas Plant Pathol 41:59–68. https://doi.org/10.1007/s13313-011-0090-6

Burkill IH (1966) A dictionary of the products of the Malay peninsula. Ministry of Agriculture and Cooperatives, Kuala Lumpur, p 2444

Davenport TL (2007) Reproductive physiology of mango. Braz J Plant Physiol 19(4):363–376

Dofuor AK, Quartey NK, Osabutey AF et al (2023) Mango anthracnose disease: the current situation and direction for future research. Front Microbiol 14:1168203. https://doi.org/10.3389/fmicb.2023.1168203

Erwin DC, Ribeiro OK (1996) *Phytophthora* diseases worldwide. The American Phytopathological Society, St Paul, p 562

FAO (2023a) Durian global trade overview. Rome

FAO (2023b) Integrated pest management. https://www.fao.org/agriculture/crops/thematic-sitemap/theme/spi/scpi-home/managing-ecosystems/integrated-pest-management/it/

Kranz J, Schmutterer H, Koch W (1977) Diseases pests and weeds in tropical crops. Verlag Paul Parey, Berlin, p 666

Mukherjee SK (1972) Origin of mango. Econ Bot 26(3):260–264

Noor Camillia NA, Salma I, Mohd Norfaizal G (2019) Development of SCAR markers for rapid identification of resistance to Phytophthora in durian using inter simple sequence repeat markers. Asian J Adv Basic Sci 7(1):30–34

Ploetz RC (ed) (2003) Diseases of tropical fruit crops. Cabi Publishing, p 481

Sapak Z, Nusaibah SA (2024) Common diseases in pineapple and their management. In: Wong MY (ed) Advances in tropical crop protection. Springer, Cham. https://doi.org/10.1007/978-3-031-59268-3_7

Singh A, Chow C, Nathaniel K, Lip Vun Y, Javad S, Jabeen K (2025) Management of *Phytophthora* and *Phytopythium* oomycete diseases in durian (*Durio zibethinus*). Crop Prot 190:107086. https://doi.org/10.1016/j.cropro.2024.107086

Soh WK (2017) Taxonomy of Syzygium. In: Nair KN (ed) The genus Syzygium: Syzygium cumini and other underutilized species. CRC Press, New York

Zakaria L (2022) Fungal and oomycete diseases of minor tropical fruit crops. Horticulturae 8(4):323. https://doi.org/10.3390/horticulturae8040323

Chapter 8
The Future: Sustainability, Food Security and Resilience in an Age of Climate Change

Abstract There are perhaps two ways to tell the future. One is to use freight train trends (Martin, Target Earth: the grand scale problems of the 21st Century, Public Lectures of the Oxford Martin School, 2008, founder of the twenty-first century School at Oxford) to shape the narrative for the future. A second way comes from the much-quoted 'The best way to predict the future is to create it', which has been attributed to many people, including the management specialist P. Drucker. I shall use the former to guide this chapter, in the hope that lessons learned will help us create our desired future.

8.1 A Tale of Freight Train Trends

Telling the future is said to be a fool's errand. That of disease control in agriculture is particularly challenging because of its dependencies on a complex and evolving dual (host and pathogen, or triple if disease vectors are included) biological system, operating in a rapidly changing physical and social environment. For example, as we speak today, the world is gripped by a climate that is heating up faster than expected, whilst decision-makers are often embroiled in political, economic, social, and technological arguments on the best way forward. The potential impact of climate change on plant diseases has already been reviewed by several authors for more than two decades now (e.g. Chakraborty et al. 2000; Jeger et al. 2023). Modelling and case studies (e.g. Chakraborty and Newton 2011) can rightly help to prepare for the future, but the complexity of future interactions means that answering the question 'what we know we do not know' may not be enough. We also need to explore 'what we do not know we do not know'.

Perhaps it is easier within the context of this book, to start by discussing 'freight train' trends (trends that cannot quickly change speed or direction, Martin (2008)), the effects of which are already apparent to us today. For example, if the present climatic trends continue, then it is obvious that with increasing temperatures it would be possible to grow tropical crops where it was before excluded because of insufficient 'growing degree days' (accumulated heat available for growth, EEA 2021).

Experience in the past (e.g. see Chap. 1 and that on plantation crops) shows that growing crops in new areas can have very positive effects on the economy of the new region, provided plant quarantine/phytosanitary measures, where appropriate, are applied.

On the other hand, it is also possible that certain crops would become more difficult to grow in existing areas as temperatures rise and water becomes limiting because of increasing drought. In other areas excessive amounts of precipitation and waterlogging of soils would favour diseases like those caused by *Phytophthora*, and demand new crop and agronomic adaptations. Whilst most attention in soybeans has been focused on rust for the last two decades (Chap. 4), stem canker and seed rot (caused by a *Diaporthe/Phomopsis* complex) are an emergent risk during recent periods of extended wet weather (Mena et al. 2023; Dos Santos et al. 2024). The corollary of this expectation is that the size and species composition of pest and pathogen populations would shift in response to selection for adaptation to the changes in the environment and host. New crop genotypes that thrive best in these new conditions will also undoubtedly emerge.

New threats will emerge as pathogens find niches that favour their reproduction, requiring alert systems of phytosanitation and quarantine. *Xylella fastidiosa* originally from coffee plants in South America, was only discovered on olive trees in Italy in 2013, but has since caused havoc with this key crop in the southern regions of the country (Saponari et al. 2019). Existing pathogens on old crops could emerge on new hosts (see example of the increasing incidence of *Magnaporthe oryzae* on wheat, Chap. 3). It may also be that pathogen populations that move to new areas will behave very differently from those that we are used to in their parent populations ('founder principle') and therefore require different approaches for their management.

8.2 Studying the Impact of Climate Change

Based on multiple regression analysis data from 1980 to 2010, Sakar et al. (2020) found a negative and significant relationship between annual average temperature and oil palm production. Modelling work by Paterson (2019) also indicated that levels of basal stem rot of oil palm are expected to increase with projected climate changes in Indonesia.

Ghini et al. (2011) summarized literature from ten reports on the expected effects of climate change on ten tropical pathosystems of tropical crops. Citing these reports, they concluded that, due to expected temperature rises, areas suitable for coffee production are likely to be significantly reduced at a number of locations in Mexico and Brazil. Losses due to coffee leaf rust are likely to increase because the latent periods of the pathogen decrease as CO_2 levels rise.

They considered that black sigatoka of banana is likely to decrease as a result of a reduction in relative humidity, while Panama disease will increase because of host stresses associated with rising temperatures and increasing drought. In common with other authors like Chakraborty and Newton (2011), they acknowledged the limitations of their conclusions because the majority of cases referred to were based on modelling rather than empirical data. It was also not always possible to draw general conclusions because of the complex number of variables.

This conclusion is hardly surprising because, whilst the classic disease triangle consists of only three interacting components (host, pathogen and environment, four if you add vectors), each component can be divided into further sub-components. As a simple example, the environment could be divided into a number of physical and biological factors, each of which is, in turn, potentially multifaceted.

8.2.1 The Complexity Theory Approach

An alternative approach is suggested by complexity theory. It would be beyond the scope of this book to discuss this approach in detail, but a simplified description of Snowden's 'Cynefin' knowledge framework (Snowden and Boone 2007; Snowden and Goh 2020) may be helpful to make the point. The theory is based simply on two domains of knowledge: that which is ordered (i.e. where cause and effect operate, i.e. linear causality) and that which is not. In the unordered, complex domain, there may be too many variables to model. The simplest metaphor to explain the difference (from Snowden, pers. comm.) is to compare trying to reproduce an *exact* replica of an aircraft (which is always possible, provided structural plans are available) with that of a garden, which is nearly impossible because of the almost infinite number of biological and physical variables involved.

Therefore, with complex systems, instead of trying to model all the variables, one could instead adopt the empirical approach of probing the system by doing limited experiments, and then, when recognizable patterns appear, act to support positive outcomes and mitigate against negative ones. Note that in complex systems, no generalizations may be possible because there is no linear causality. It is, therefore, more of a heuristic rather than a rule-based approach. Time and timing are of the essence here because it is critical to catch patterns at an early stage of formation, so that pre-emptive action can be taken in time.

As an example, if one does not know the impact of climate trends on a particular crop and its pathosystem, one could undertake limited experiments based on likely changes in the environment acting on predominant hosts and pathotypes. The results may not immediately lead to generalized conclusions, but the response of the pathosystem could stimulate new perspectives or indicate what we need to pre-emptively do. When the results make sense, then we are back in the ordered domain and can return to a generalized modelling of the process.

8.2.2 Geographical Implications

Other authors have broadly argued that tropical regions are at greater risk than temperate regions from climate change. Part of the evidence suggesting this comes from an analysis by Locosselli et al. (2020) of 209 studies on the growth rings of trees in temperate and tropical forest trees. They noted that tropical trees, in general, grow on average twice as fast as those in temperate zones but live shorter lives. Mean tree longevity was higher in moist compared to dry tropical biomes. Their analysis suggests that with drier and warmer climates overall tree longevity in the tropical lowlands will be reduced. As a result, the biomass stored in these lowland forests could be significantly reduced.

8.3 Sustainability

These new challenges emerge in a world that is already at risk of food shortages, partly as a result of environmental damage due to our current systems of food production, as well as problems related to unequal distribution. Sustainability is therefore a key risk, whatever the changes the future may bring. In recent times, a debate has therefore arisen about what is sustainable from an environmental perspective and what is sufficiently productive to meet our growing food needs.

This has led to a plethora of sustainability concepts intended to minimize the negative impacts of agriculture on the environment and to regenerate lost resources. These range from exclusively 'organic' methods of cultivation (with complete abstinence from the use of 'synthetic' pesticides, fertilizers and genetically modified organisms), and those that focus exclusively on soil health, to those that advocate sustainable forms of conventional agriculture (i.e. usually large-scale, mechanized, open to the best technological options, but keeping environmental considerations paramount).

Oberc and Schnell (2020) noted 14 approaches that are aligned with the concept of Regenerative Agriculture:

Agroecology.
Nature-inclusive agriculture.
Permaculture.
Biodynamic agriculture.
Organic farming.
Conservation agriculture.
Regenerative agriculture.
Carbon farming.
Climate-smart agriculture.
High nature value farming.
Low external input agriculture.

8.3 Sustainability

Circular agriculture.
Ecological intensification.
Sustainable intensification.

For simplicity I shall here use the term Regenerative Agriculture (RA) as defined by Oberc and Schnell (2020) and EASAC (2022): 'a system of farming principles that aims to maintain agricultural productivity, increase biodiversity and in particular restore and maintain soil biodiversity, and enhance ecosystem services including carbon capture and storage' whilst maintaining productivity (see also Newton et al. 2020; Giller et al. 2021). In contrast to the other related concepts, regenerative agriculture is not viewed as defined *a priori* by a given set of rules and practices. Instead, the goals that should be achieved are set. 'Practices and new technologies are adopted over time which contribute to achieve these goals' (EASAC 2022).

Central to the majority of these approaches to RA is the need to support ecosystem services, which include soil fertility and health, hydrological considerations (e.g. water availability, quality, flood control), a biodiversity of crops (e.g. mixtures and agroforestry, rotations) and of self-limiting predator-parasite populations (e.g. where the predominance of damaging genotypes can be limited by frequency-dependent selection) and the health of pollinating agents.

Measures to support climate change mitigation would also require C sequestration (e.g. in integrated systems of livestock-crop production).

A story I first heard during an international plant pathology congress in Latin America some years ago will perhaps serve to exemplify some of these complex considerations. The principles are elaborated in the thought-provoking paper on 'deforestation and reforestation' by Ruf and Zadi of CIRAD (1998).

8.3.1 Storypad 16: Mario's Tale

Mario (a fictitious name) used to work for a logging company. The land he had helped to clear was cheap, had rich, fertile soil and a good water supply. There was little evidence of the potential for disease and pest problems from nearby farms. So, with the money he saved from his former job, Mario decided to invest in a piece of land and raise a family on it.

According to what he had heard in the city about product prices, perennial plantation crops could be a good money earner over the long term.

Sure, there was a 4–5 year lag time between planting a crop till it came to maturity, but Mario was resourceful and grew short-term crops and reared livestock to supplement his income during this time. He grew a mixture of crops, and the livestock provided fertilizer. Even the bananas and coconut trees he planted to shade his newly planted perennial crop were a source of income. As a bachelor, his daily expenses were low.

In the sixth year, the perennial crop started to bear fruit. His income shot up, and he could now afford to start a family. With their help and some paid workers, he initially managed his trees well, mulching and cover cropping his land regularly. His income was now at its highest over the next 25–30 years. As a plantation owner of a valuable cash crop, he became well-known for his generous contributions to the welfare of his village and could even afford to send his children to school in the city.

From the 30th year onward, however, the yields of his aging crop began to decline. The years had taken their toll, and he now lacked the energy to personally supervise the care and management of his trees in the manner he used to do in earlier years. His children were now professionals in the city and had no interest in returning to rural life. He was still expected to contribute to the next wedding in the village. At this time, pest and disease problems also increased in extent and severity. Because there was little genetic variability among his trees, pathogens spread unimpeded. Water for irrigation became harder to access.

Mario was now faced with a dilemma. If he were to replant, it would now be costly. It would be difficult to restore and maintain the natural soil fertility he had enjoyed over the years. The cost of acquiring new land would be prohibitive, and given the emergence of new pests and diseases, there was no guarantee that he would be able to enjoy the fruits. Water had become limiting, requiring ever deeper wells. Should he cut his losses, pack up, and emigrate to another location/country?

In short, the benefits of his 'forest rental' (Ruf et al. 1995), e.g. rich, healthy soil, good hydrological balance, and freedom from pest and disease epidemics, had run out. He could either try to restore forest rental (revert to forest) or replant in a new area. Neither seemed feasible at his age.

He wondered what difference it would have made if he had introduced some intervention points in the past years (e.g. introduced better soil protection, crop husbandry and IPM measures; systematically rejuvenated the plantation by periodic replacement of poor/aging trees; as well as crop diversification), as his local agricultural agent had suggested?

The story speaks to us of the central role of conserving ecosystem services that enable sustainability in the production of crops in the future. We have already discussed the reasons in earlier chapters why the tropics are particularly vulnerable to environmental change. Ironically, while tropical rainforests have the highest biomass among all forms of vegetation, their soils can be relatively nutrient-poor, and organic matter in the topsoil needs to be replenished regularly due to its rapid decomposition under conditions of high temperatures and humidity (Tho 1987). It has been noted before in earlier chapters that the tropics do not enjoy the benefit of a cold season, where the cyclic growth of pests and diseases can be interrupted.

It also speaks to us about the value of biodiversity in crops, as exemplified by the stability in Mario's initial (smallholder) system of crop production. Due to their great biodiversity, virgin tropical forests are typically highly stable ecosystems. On the other hand, monocropping on large areas of land is common in modern, conventional agriculture. As a result of the loss of within-crop biodiversity, such agroecosystems often lack resilience and need to be carefully managed.

8.4 Resilience

The value of biodiversity in strengthening crop resilience has been demonstrated in barley by Wolfe and Barrett (1980), in agroforestry by Wolfe (2019), in field experiments on rice (Chin and Ajimilah 1982), and in production areas of rice (Zhu et al. 2000). More recently, the increased incidence of Goss's wilt disease of corn (caused by *Clavibacter michiganensis* subsp. *Nebraskensis*) in the United States is thought to have been aggravated by alternative hosts serving as refuges and by reduced crop rotation (Campbell et al. 2019).

The rationale behind the use of diversity in crops (in space or in time) is simple. Crops that contain mixtures of cultivars (cultivar mixtures) or breeding lines ('multilines') impose a multidirectional selection pressure on the pathogen population. This potentially disruptive selection reduces the speed of adaptation (virulence development) by the pathogen to any one host resistance. Indeed, the genetic penalty that is extracted in overcoming these host systems could be so high that it is beyond the evolutionary capacity of the pathogen population to easily overcome. The end result is greater durability of host resistance to disease. To illustrate the problem posed to pathogens and pests, Martin Wolfe used to cite the simple example of the difficulties in breeding a dog that is both a strong watch dog and a speedy greyhound.

There are also physiological and agronomic advantages to crop diversity. For example, it can enable better exploitation of the growing environment (e.g. varying rooting patterns that better exploit nutrients and water available in the soil) and produce 'induced R' responses in the host due to multiple, heterogeneous infections (Chin et al. 1984).

In the same manner, mixtures of crop protection agents can make it more difficult for pathogens to develop resistance to any one of these agents (e.g. Chin and Holloman 2000). These mixtures could be prepacked, or individual agents could be applied in rotation (i.e. time-based diversification, e.g. in bananas) or in different fields (spatial diversification). Taken to the next level of sophistication, the combination of cultivar diversification with that of targeted control agents offers a further increase in the complexity of the genetic challenge presented to pathogens and pests.

In the broader context of crop and pest/pathogen co-evolution in agriculture, it seems likely that many of the arable pests and diseases we encounter today would not have achieved the prominence they have without encountering domesticated crops and their production in a man-made environment. For example, arable weed species related to specific crops may be reliant on cultivated fields as a habitat. Interestingly, recent concepts of weed management in crops have argued for maintaining weed species biodiversity to support agroecosystem services (Gaba et al., 2016).

8.5 IPM

The integration of *all* available tools in a sustainable manner leads us to Integrated Pest Management (IPM). The European Commission defines IPM to mean 'a careful consideration of all available plant protection methods and subsequent integration of appropriate measures that discourage the development of populations of harmful organisms, keep the use of plant protection products and other forms of intervention to levels that are economically and ecologically justified, and reduce or minimize risks to human health and the environment. "Integrated pest management" emphasizes the growth of a healthy crop with the least possible disruption to agro-ecosystems and encourages natural pest control mechanisms.'

In the EU, IPM has become a cornerstone of the Directive on the Sustainable Use of Pesticides (European Parliament and Council 2009), and in the United States, it is promoted under the Pesticide Environmental Stewardship Program (PESP). The FAO promotes the tool as the preferred approach to crop protection, with the FAO IPM program currently comprising three regional programs in Asia, the Near East and West Africa (FAO 2023). The integration of new technologies, such as novel bioactives, precision delivery of control agents, biostimulants and other biocontrol agents, as well as diagnostics and a better understanding of the genetic structure and behaviour of pest and pathogen populations will significantly improve our toolset to deal with the challenges ahead.

As a classic example of IPM, sigatoka control in banana plantations involves cultural methods like de-leafing of infected leaves to reduce inoculum pressure; soil additives to accelerate the decomposition of dead leaves on the ground; maintenance of plant health through appropriate fertilizer application, weed and nematode control; good soil drainage and drip irrigation to minimize air humidity (Marin et al. 2003). The application of fungicides, supported by the careful monitoring of disease (e.g. 'youngest leaf streaked') has been described earlier in the present text. Doubtless, host resistance will become another important tool in the future.

Whilst diversity would appear to be a panacea for strengthening resilience in agroecosystems, in practice a balance is needed between productivity and resilience. As noted earlier, EASAC (2022) defines Regenerative Agriculture to include maintaining productivity. Two tales (in italics) from complexity/systems studies (Longstaff 2011) offer food for thought. My comments are in upright text.

8.5.1 Storypad 17: Food for Thought from 'Systems Studies'

Castles are good as long as the walls are not breached (Longstaff 2011).

Similar catastrophic consequences can occur when single powerful R genes/chemicals/control measures are overcome by genetic adaptation of the pathogen. In tropical crops, the rapid development of fungicide resistance in Black Sigatoka to the benzimidazole (in some cases, e.g. after only 3 years of use in Honduras) is a well-known story. It ultimately resulted in the withdrawal of the fungicide from use on bananas in some South American countries (Marin et al. 2003). Similar observations can be made about benzimidazole on powdery mildew of cucurbits (McGrath 2001).

In a broader sense, the trend towards single-site products in crop protection could be thought of as a trade-off. For example, the activity and selectivity of these modern fungicides make them safer to use at much lower application rates of active ingredient per ha of crop. On the other hand, the specific sites of activity have exposed them to simple genetic adaptation by pathogens. Interestingly, unlike the majority of single-site products today, none of the older multisite fungicides (see table on 'Agricultural fungicides used on tropical crops' in Appendix 1 of this book) have required the use of anti-resistance strategies. Indeed, they have been used as mixing partners with single-site products to extend their activity.

> Bundles of sticks are stronger than the combined strength of individual sticks, except when there is a fire. All sticks in a bundle will be burnt, but at least some will survive if they are not bound together (Longstaff 2011).

It is true that diversification of crop stands can improve resilience, in that negative effects on any one component can be compensated by growth in another component, thereby allowing the system as a whole to 'bounce back'. Several other advantages of variety mixtures of cereals have been described by Chin and Wolf (1984). These abilities depend, however, on how tightly the components are coupled together, and if the coupling has disadvantages. There could, for example, be unintended consequences of mixtures of crops where the final product is too heterogeneous to meet market standards or infringes on breeders' rights to individual cutivars.

Resilience has many other facets in crops. The ability of some crops to withstand climatic change, environmental extremes, pest attack and feeding, even man-made catastrophes (like conflict and arson), and yet bounce back is exemplified by cassava. The tuberous roots are protected and store-well deep underground. They are tolerant of drought, pests and diseases and quickly regenerate new plants. Adaptability is a different, but related concept, exemplified perhaps by deep-water rice that keeps on increasing in height to keep its head above rising floodwaters (Chap. 3).

Durians are especially adapted to survive under forest conditions where there may be a wide range of frugivores (fruit eaters). It produces a highly aromatic, pulpy aril around its seeds to attract animals but does not want them to be consumed by small animals, which would restrict its area of dispersion from the parent tree. The solution is apparently to produce large, heavy, spiky fruits, with a thick husk. The fruit, which looks like a medieval mace, is virtually impregnable until it is ripe (a property I have foolishly tried before to demonstrate before a class of undergraduates).

8.6 Technological Advances

Advances in diagnostics and surveillance technologies have been discussed (see also Storypad 11; Jeger et al. 2023) as vital components of the future management of diseases. This is particularly true for diseases that are cryptic (hidden), like those caused by viruses and phytoplasmas (e.g. in palm trees), or are not easily visible (like VSD disease of cocoa, *Fusarium* of several crops) or difficult to locate on large trees (e.g. early stages of basal stem rot of oil palm). The use of PCR and related methods like LAMP will enable more effective methods of disease management through speedier and more precise field diagnosis (Adusei-Fosu and Dickinson 2023). Aerial monitoring of virulence (to host resistance) or resistance (to chemical agents) can also be supported by PCR methods. The use of spore traps and 'mobile nurseries' to monitor aerial inoculum were useful methods in the past, which can be updated with new analytical, molecular-based technologies. Predictive mathematical modelling of disease and pest epidemics will inform on the design of longer-term preventive strategies (e.g. for cassava, Godding et al. 2023) and the improvement of AI-based decision systems for the implementation of intervention measures.

Advances in genomics and transcriptomics will speed up and bring about greater precision in the characterization of functional genetic resources of crops (e.g. in seed banks) and the breeding of improved cultivars, which were previously constrained because of genetic barriers. A commonly cited example is the problem that polyploidy can create for the hybridization of certain crops, like bananas. Despite the availability of a broad genetic diversity of thousands of local cultivars, little progress with bananas was before possible because commercial cultivars were mostly triploid and sterile (Heslop-Harrison and Schwarz 2007). New knowledge of structural and functional genomics has recently enabled progress in improving resistance to abiotic stress (e.g. drought) and diseases (e.g. *Fusarium*), desirable agronomic features (height, leaf angle and root system), as well as in nutraceuticals (Wang et al. 2021; Tripathi et al. 2023). RNA-related methods have introduced new opportunities for breeding plant resistance to viruses (Tatineni and Hein 2022).

Rice is well known as the first cereal crop to have its genome fully sequenced (Nature 2005) and has served as a model system for other cereals with larger genomes. Amongst the many applications and spin-offs of this research (Jackson 2016; Hanafiah et al. 2020) for rice improvement, it is helping to address the age-old conundrum of whether one should breed for higher tiller/panicle numbers per plant or greater weight of individual panicles (Zhao et al. 2020).

As is apparent from the present research programs of universities and reports in the literature, the genetic improvement of our crops, as reported in the literature, is increasingly based on molecular innovations. Interestingly, whilst methods are changing, the ultimate objectives remain largely embedded in productivity, resilience against biological and environmental threats, and sustainability.

Measures for direct intervention, such as the use of disease control agents, have, in addition to improved human and environmental safety, become more active at lower rates of application. Umetsu and Shirai (2020) noted, for example, that use rates of pesticides have been reduced from 1 to 10 kg/ha in the mid-1900s to 10 g/ha in a significant number of cases today. This trend is likely to continue. Systems of delivery will continue to be refined for more precise and targeted applications. Innovative new systems will be developed, e.g. recently in the use of bees ('bee vectoring') to deliver a naturally occurring microbial (*Clonostachys rosea* CR-7) to flowers for the control of *Botrytis* and *Monilinia* in strawberries (Ogden 2020).

Modern crop cultivars will undoubtedly be purpose-selected for different systems of cultivation (some of which we may not even know about today) and adapted to the environmental constraints of the future. Potentially, they could be more responsive to novel biostimulants/bioactives, for increased yield and tolerance to environmental stress or to support new defensive genetic systems. Knowledge of their genetics at a molecular level could perhaps enable them to be participants in new systems of food production itself, that are independent of the plant crops that we use today.

Advances in agricultural technology are also increasingly characterized by collaborative science networks, often multinational in reach, and enabled by a connected world. These range from the 'international testing programs' nurtured by CGIAR crop institutions, to consortia of largely voluntary groups joined by a common interest in a particular area of science. The success of the former has, for example, led to the strengthening and maturity of national science institutions in the developing world. The spectacular success of the rice genome project, which involved scientists from six countries and contributions from two industry groups, has been a testament to the application of genomics for crop improvement. Collaborative science will become an increasingly vital facet of research in the future.

8.7 Artificial Intelligence: A Second Freight Train Trend?

Perhaps a second 'freight train trend' that is recently upon us is the impact of the rapid progress of AI (Artificial Intelligence) and GenAI, which are today riding the crest of the digital revolution. They have begun to catalyze the way that machines can augment and perhaps multiply human intelligence. Mollick (2024) has described this as co-intelligence.

In agriculture, it seems likely that AI will contribute greatly:

- In R&D, by finding, characterizing, improving and testing new active ingredients and sources of genetic resistance for disease and pest control or enhanced performance of crops; accelerating diagnosis and impact assessment of pests, diseases and environmental constraints; supporting the evaluation of safety and environmental risks; and planning and assessing new strategies for their management at reduced costs and safe use.

- In the field, AI-based predictive and decision models of weather, soil, crops, agronomic and crop protection measures to optimize resource use whilst minimizing environmental damage.
- In marketing, to improve models that can help better map, predict demand and optimize supply chain operations, and manage expectations and improve communications with farm customers.
- In developing 'smart' technologies to provide a whole raft of tools and measures to improve the efficiency of farm operations, e.g. the use of robotics and automation in tillage, weed management, irrigation and in the targeted application of agronomic/crop protection/biostimulant products.

In the context of this book, GenAI could lead to behavioural changes in the way we access and use information. This suggests that the traditional didactic role of textbooks as trusted, encyclopaedic stores of library-based information will need to be adapted to do more. This book, in itself, is an experiment in collaborative and interactive science (see Epilogue) to encourage self-learning, discovery and experience sharing. I therefore thank the reader for participating in this study.

References

Adusei-Fosu K, Dickinson M (2023) Detecting Fusarium oxysporum f. sp. elaeidis by using loop-mediated isothermal amplification. J Plant Pathol 105:1637–1643. https://doi.org/10.1007/s42161-023-01435-9

Campbell TM, Ikley JT, Johnson WQG, Wise KA (2019) Impact of inoculum concentration on Goss's wilt development in corn and alternative hosts. Plant Health Progress 20:155–159. https://doi.org/10.1094/PHP-01-19-0002-R

Chakraborty S, Newton AC (2011) Climate change, plant diseases and food security: an overview. Plant Pathol 60:2–14

Chakraborty S, Tiedemann AV, Teng PS (2000) Climate change: potential impact on plant diseases. Environ Pollut 108(2000):317–326

Chin KM, Ajimilah NH (1982) Rice variety mixtures in disease control. In: Proceedings of the international conference of plant protection in the tropics, Kuala Lumpur, pp 241–246

Chin KM, Holloman DW (2000) Fungicide resistance risk management. In: Proceedings BCPC symposium, pp 31–36

Chin KM, Wolf MS (1984) The spread of *Erysiphe graminis* f.sp *hordei* in mixtures of barley varieties. Plant Pathol 33:89–100

Chin KM, Wolfe MS, Minchin P (1984) Host mediated interactions between pathogen genotypes. Plant Pathol 33:161–171

Dos Santos GC, Lima Horn LM, Trezzi Casa R, Soardi K, Lopes MA, Nascimento SCD, Santi VM et al (2024) First report of seed decay caused by Diaporthe ueckeri on soybean in Brazil. Plant Dis. https://doi.org/10.1094/PDIS-04-24-0814-PDN

EASAC (2022) Regenerative agriculture in Europe, EASAC policy report, p 58

EEA (2021) Growing degree days. https://www.eea.europa.eu/data-and-maps/figures/growing-degree-days

European Parliament and Council (2009). Integrated Pest Management, Directive on the Sustainable Use of Pesticides. https://food.ec.europa.eu/plants/pesticides/sustainable-use-pesticides/integrated-pest-managementipm_en#Related

References

FAO (2023) Integrated pest management. https://www.fao.org/agriculture/crops/thematic-sitemap/theme/spi/scpi-home/managing-ecosystems/integrated-pest-management/it/

Gaba S, Rebouda X, Fried G (2016) Agroecology and conservation of weed diversity in agricultural lands. Bot Lett 163(4):351–354. https://doi.org/10.1080/23818107.2016.1236290

Ghini R, Bettiol W, Hamada E (2011) Diseases in tropical and plantation crops as affected by climate changes: current knowledge and perspectives. Plant Pathol 60(1):122–132

Giller KE, Hijbeek R, Andersson JA, Sumberg J (2021) Regenerative Agriculture: an agronomic perspective. Outlook Agric 32(1):13–25

Godding D, Stutt ROJH, Alicai T, Abidrabo P et al. (2023) Developing a predictive model for an emerging epidemic on cassava in sub-Saharan Africa. http://www.nature.com/scientificreports

Hanafiah NM, Mispan MS, Lim PE, Baisakh N, Cheng A (2020) The 21st century agriculture: when rice research draws attention to climate variability and how weedy rice and underutilized grains come in Handy. Plan Theory 9:365. https://doi.org/10.3390/plants9030365

Heslop-Harrison JS, Schwarzacher T (2007) Domestication, genomics and the future for Banana. Ann Bot 100(5):1073–1084

Jackson SA (2016) Rice: the first crop genome. Rice 9:14. https://doi.org/10.1186/s12284-016-0087-4

Jeger MJ, Fereres A, Malmstrom CE, Mauck KE, Wintermantel WM (2023) Epidemiology and management of plant viruses under a changing climate. Phytopathology 113:1620. https://doi.org/10.1094/PHYTO-07-23-0262-V

Locosselli GM, Brienen RJW, Leite MS, Gloor M, Krottenthaler S, Oliveira AA, Barichivich J, Anhuf D, Ceccantini G, Schöngart J, Buckeridge M (2020) Global tree-ring analysis reveals rapid decrease in tropical tree longevity with temperature. Proc Natl Acad Sci USA 117(52):33358–33364. https://doi.org/10.1073/pnas.2003873117

Longstaff PH (2011) Resilience in systems. dealing with the new normal. Innovation brown bag seminars, knowledge management, ETH University of Zurich

Marín DH, Romero RA, Guzmán M, Sutton TB (2003) Black Sigatoka an increasing threat to banana cultivation. Plant Dis 87(3):208–222

Martin J (2008) Target Earth: the grand scale problems of the 21st Century, Public Lectures of the Oxford Martin School. University of Oxford. https://www.oxfordmartin.ox.ac.uk/events/public-lecture-target-earth-by-james-martin

McGrath MT (2001) Fungicide resistance in cucurbit powdery mildew. Plant Dis 85(3):236–245

Mena E, Stewart S, Montesano M, de Leon IP (2023) Plant Pathol 73(1):31–46. https://doi.org/10.1111/ppa.13803

Mollick E (2024) Co-intelligence: living and working with AI, Penguin, p 256

Nature (2005) International Rice genome sequencing project and Sasaki T. The map-based sequence of the rice genome. Nature 436:793–800. https://doi.org/10.1038/nature03895

Newton P, Civita N, Frankel-Goldwater L, Bartel K, Colleen JC (2020) What is regenerative agriculture? Front Sustain Food Syst 4:1–11

Oberč BP, Schnell A (2020) Approaches to sustainable agriculture. Exploring the pathways towards the future of farming. IUCN EURO, Brussels

Ogden LE (2020) Biocontrol 2.0: a shifting risk–benefit balance. Bioscience 70:17–22. https://doi.org/10.1093/biosci/biz135

Patterson RRM (2019) Ganoderma boninense disease of oil palm to significantly reduce production after 2050 in Sumatra if projected climate change occurs. Microorganisms 7:24. https://doi.org/10.3390/microorganisms7010024

Ruf F, Zadi H (1998) Cocoa: from deforestation to reforestation. In: First international workshop on sustainable cocoa growing. Smithsonian Tropical Research Institute, Panama City

Ruf F, Konan G, Ardhy W (1995) Forest rent, replanting and regulation of cocoa supply. In: Conference paper, 8th meeting of the advisory group on the world cocoa economy, Yaounde

Saponari M, Giampetruzzi A, Loconsole G, Boscia D, Saldarelli P (2019) Xylella fastidiosa in olive in Apulia: where we stand. Phytopathology 109:175–186. https://doi.org/10.1094/PHYTO-08-18-0319-FI

Sarkar MSK, Begum RA, Pereira JJ (2020) 2020 impacts of climate change on oil palm production in Malaysia. Environ Sci Pollut Res 27:9760–9770. https://doi.org/10.1007/s11356-020-07601-1

Snowden DJ, Boone ME (2007) A leader's framework for decision making. Harv Bus Rev 85(11):68–76

Snowden DJ, Goh Z (2020) Cynefin – weaving sense-making into the fabric of our world. In: Riva Greenberg, Boudewijn Bertsch (eds), The Cynefin Co, p 376

Tatineni S, Hein GL (2022) Plant viruses of agricultural importance: current and future perspectives. Phytopathology 113:117–141. https://doi.org/10.1094/PHYTO-05-22-0167-RVW

Tho YP (1987) Ecological and biological considerations in the management of the tropical rainforest ecosystem. In AMIC-CDG-COMCON-UKM Workshop on Mass Media and the Protection of the Environment: Kuala Lumpur, Asian Mass Communication Research and Information Centre, Singapore. https://hdl.handle.net/10356/101056

Tripathi JN, Ntui VO, Malarvizhi M, Muiruri S, Ravishankar KV, Tripathi L (2023) Improvement of nutraceutical traits of Banana: new breeding techniques. In: Kole C (ed) Compendium of crop genome designing for nutraceuticals. Springer, Singapore

Umetsu N, Shirai Y (2020) Development of novel pesticides in the 21st century. J Pestic Sci 45(2):54–74. https://doi.org/10.1584/jpestics.D20-201

Wang X, Yu R, Li J (2021) Using genetic engineering techniques to develop Banana cultivars with Fusarium wilt resistance and ideal plant architecture. Front Plant Sci 11:617528. https://doi.org/10.3389/fpls.2020.617528

Wolfe MS (2019) Agroforestry can decentralise production of food and energy and encourage 'commoning'. In: 4th world agroforestry congress, Montpellier

Wolfe MS, Barrett JA (1980) Can we lead the pathogen astray? Plant Dis 64:148–155

Zhao S, Jang S, Lee YK, Kim DG, Jin Z, Koh HJ (2020) Genetic basis of tiller dynamics of rice revealed by genome-wide association studies. Plants (Basel) 9(12):1695. https://doi.org/10.3390/plants9121695

Zhu Y, Chen H, Fan J, Wang Y, Li Y, Chen J, Fan J, Yang S, Hu L, Leung H, Mew TW, Teng PS et al (2000) Genetic diversity and disease control in rice. Nature 406(6797):718–722. https://doi.org/10.1038/35021046

Chapter 9
Epilogue: Weaving the Narratives Together

Abstract With the reader's indulgence, I would like to try and end as most good stories do, with a recount of our journey and a slight 'twist' to remind the reader of the purpose of this book.

The Storypads and snippets distributed throughout the text, as well as the facts assembled in the appendices, are my own take on what I thought I witnessed on our journey. They are distributed according to related crop groupings, but they also tell an overall tale of biodiversity, sustainability, resilience and adaptability.

We began by outlining some examples of how plant diseases have helped shape and continue to shape human affairs (Storypad 1), and examined, using rice as an example, the genetic variability that plants have evolved to deal with the challenges of a diverse and changing environment (Storypad 2). We touched on areas that we do not normally appreciate, for example, the elegant images and adaptations of organisms (Storypad 3) that cause our crops so much damage; or the microscopic pest that is representative of the most abundant form of animal life on this planet (Storypad 14). Microorganisms are well known to be useful sources of food and industrial chemicals, but pathogens are not normally thought of as a food delicacy (Storypad 4) or panacea (Storypad 11). How the economic success of communities and cultures can be intricately linked with crops and diseases is the topic of Storypad 5. As the world is increasingly threatened today by extreme events of climate and geology and human conflict, it is insightful to think of how plants have evolved special mechanisms for survival (Storypad 6). Storypad 7 elaborates on how the potential of a crop depends on how well it meets the needs of the challenges of today. Storypad 8 provides two examples of how pathogens can influence our 'favourite' choice of fruits. Apart from their popular domestic applications in the tropics, the large leaves of bananas have a surprising use for research: they serve as microplots for field experiments (Storypad 10). Storypad 14 relates a tale about two legends of tropical crop protection in SE Asia, a crop and a disease. Molecular methods are revolutionizing our crop protection toolset. For example, many diseases that were cryptic and difficult to diagnose for mitigatory measures can now be monitored 'just-in-time' using PCR-related methods (Storypad 12), and genomics has, of course, greatly assisted progress in the genetic improvement of crops (see this chapter). The integrated approach of combining all effective control and agronomic measures to manage disease (IPM) has long been endorsed by the EU, USDA and FAO. One man's daily routine (Storypad 9) is an excellent example of the IPM that growers adopt naturally without being prompted to do so by science. It also reminds us of the complex management decisions our growers make to bring food to our table. Future food production can only be secure if it is sustainable, resilient and adaptable. Storypad 16 speaks to us about one man's experience with sustainability based on ecosystem services. Storypad 17 challenges us to explore studies in other disciplines for new insights into sustainability and resilience in agriculture.

But my own perspectives are not the essence of this book. My recollection of narratives and views, factual trends (in appendices) and extensive citations of other people's work are not, in themselves, the main reason for putting this work together.

Irrespective of whether you agree or disagree with my views, much more important are potentially your *own* insights, which may have been triggered simply by engaging with the narratives. Experience suggests that the more violently you feel about the treatment of these topics, the more likely it is that the process could have generated new perspectives that are unique to you, or reminded you of your own, perhaps long 'forgotten' stories.

The French scientist and philosopher Blaise Pascal (1670) wrote about the benefits of self-discovery: 'People are generally better persuaded by the reasons which they have themselves discovered than by those which have come into the mind of others'.

In this manner, I hope that the process has not only reminded you of what you perhaps already know but also helped you know a little more of 'what you did not before know, that you did not know'. If one accepts lessons from the past, which show that the most compelling 'black swan' events were those that were the least expected, then answering this question could be a key part of anticipating the complex challenge of scientific research for the future.

Postscript

Interestingly, students in international agriculture at HAFL (University of Bern, CH) have now, for a number of years, been trained using a not dissimilar process called 'problem-based learning', in which there are no formal lectures, but students are instead guided to solve stories of situational problems by researching them on their own (Prof. Scheidegger U.C. *pers. comm.* 2024).

Recent progress in GenAI also emphasizes the importance of moving textbooks beyond the role of just providing encyclopaedic information to that of actively facilitating learning and discovery.

Appendices

Appendix 1

Chemical and Biological Control Agents of Diseases in Tropical Crops

The table is largely organized around chemical and non-chemical groupings that are recommended by FRAC (2024). It includes single- and multisite, organic and inorganic fungicides, microbial and chemical host defense inducers and biocontrol products. IPM and vector control examples have been added. Tropical crop diseases and related control agents are individually referenced. Entries are not meant to be exhaustive, but only as examples of control agents on tropical crops as reported in the scientific literature. No commercial names are mentioned.

Extended reading: common names of chemicals and other details may be found on the FRAC website or in the Pesticide Properties Database of the University of Hertfordshire, United Kingdom (https://sitem.herts.ac.uk/aeru/ppdb/).

Fungicide group (ex FRAC)	Site of action/ multisite	Diseases (ca, causal agent)	References
Qo Inhibitor (QoI) and Qi inhibitor (QiI)	Act at the Quinone 'outer' (Qo) or Quinone 'inside' (Qi) binding site of the cytochrome bc1 complex	Asian rust of soybeans (ca *Phakopsora*)	Godoy et al. (2016)
		Blast disease of rice (ca *Magnaporthe*)	Ogoshi et al. (2018)
		Downy mildew of cucurbits (ca *Pseudoperonospora*)	Theerthagiri et al. (2008)

(continued)

© The Editor(s) (if applicable) and The Author(s), under exclusive license to Springer Nature Switzerland AG 2025
K. Chin, *Tales of tropical plant diseases in an age of climate change*,
https://doi.org/10.1007/978-3-031-90790-6

Fungicide group (ex FRAC)	Site of action/ multisite	Diseases (ca, causal agent)	References
		Early blight of tomato (ca *Alternaria*)	Jacobelis et al. (2023)
		Leaf spot of peanuts (ca *Cercospora*)	Culbreath et al. (2002, 2008)
		Patch canker and other *Phytophthora* diseases of durian	Kongtragoul et al. (2021)
		Powdery mildew of cucurbits (ca *Erysiphe*)	Theerthagiri et al. (2008)
		Sheath bight of rice (ca *Thanatephorus*)	Uppala and Zhou (2018)
		Sigatoka of bananas (ca *Mycosphaerella*)	Chin et al. (2001), Pérez et al. (2002), Marin et al. (2003)
Sterol Biosynthesis Inhibitors (SBI's)	Demethylation inhibitors (DMI's)	Asian soybean rust (ca *Phakopsora*)	Godoy et al. (2016)
		Anthracnose of mango and chili (ca *Colletotrichum*)	Nordmeyer et al. (1994)
		Basal stem rot of oil palm (ca *Ganoderma*)	Nur-Rashyeda et al. (2022) (hexaconazole by trunk injection)
		Early blight of tomato (ca *Alternaria*)	Nordmeyer et al. (1994)
		Leaf spot of peanuts (ca *Cercospora*)	Culbreath et al. (2008)
		Narrow brown leaf spot of rice (ca *Cercospora*)	Uppala and Zhou (2018)
		Powdery mildew of cucurbits (ca *Podosphaera*)	McGrath (2001)
		Sheath blight and brown spot of rice (ca *Thanatephorus* and *Cochliobolus*)	Nordmeyer et al. (1994), Gao et al. (2022)
		Sigatoka of banana (ca *Mycosphaerella*)	Chin et al. (1996), Marin et al. (2003)
		Soybean root rot (ca *Rhizoctonia*)	Ajayi-Oyetunde and Bradley (2017)
		Vascular streak dieback of cocoa (ca *Oncobasidium*)	Nordmeyer et al. (1994)

(continued)

Fungicide group (ex FRAC)	Site of action/ multisite	Diseases (ca, causal agent)	References
	Amines *(aka morpholines)*	Powdery mildew of cucurbits (ca *Podosphaera*)	McGrath and Fox (2010)
		Sigatoka of banana (ca *Mycosphaerella*)	Marin et al. (2003), Sánchez et al. (2017)
Succinate dehydrogenase Inhibitors (SDHI), aka carboxamides	Complex II in the mitochondrial respiration chain: no Oomycete activity	Asian soybean rust (ca *Phakopsora*)	Guicherit et al. (2014), Godoy et al. (2016), Claus et al. (2022)
		Early blight of tomato (ca *Alternaria*)	Sierotsky and Scalliet (2013)
		Gummy stem blight of cucurbits (ca *Didymella*)	Guicherit et al. (2014)
		Powdery mildew of cucurbits (ca *Podosphaera*)	McGrath (2001), Sierotsky and Scalliet (2013)
		Root rot of soybeans (ca *Rhizoctonia*)	Ajayi-Oyetunde and Bradley (2017)
		Sigatoka of banana (ca *Mycosphaerella*)	Silva et al. (2024)
		sheath bight of rice (ca *Thanatephorus*)	Zhao et al. (2022)
Carboxylic Acid Amides	Inhibition of cellulose synthesis in Oomycete plant pathogens	Late blight of tomato (ca *Phytophthora*)	Fanigliulo and Sacchetti (2009)
		Downy mildew of cucurbits (ca *Pseudoperonospora*)	Ojiambo et al. (2010)
		Root rot of avocado (ca *Phytopthora*)	Belisle et al. (2019)
		Patch canker and other *Phytophthora* diseases of durian	Kongtragoul et al. (2021)
Oxysterol binding proteins Inhibitors (OSBPI's)	Inhibits an oxysterol binding protein (OSBP) homologue	Black shank of tobacco (ca *Phytophthora*)	Miao et al. (2016)
		Downy mildew of cucurbits (ca *Pseudoperonospora*)	Miao et al. (2016)
		Damping off and root rot of several crops (ca *Pythium*)	Miao et al. (2016)

(continued)

Fungicide group (ex FRAC)	Site of action/ multisite	Diseases (ca, causal agent)	References
Phenylamides (PAs)	Inhibition of rRNA biosynthesis (polymerase complex I)	Late blight of tomato (ca *Phytopthora*)	Müller and Gisi (2011), Leadbeater (2014), Gisi and Sierotsky (2015)
		Patch canker and other *Phytopthora* diseases of durian	Kongtragoul et al. (2021)
		Downy mildew of cucurbits (ca *Pseudoperonospora*)	Ojimbo et al. (2010)
		Damping-off and root rot on several vegetable crops (ca *Pythium*)	Leadbeater (2014)
		Root rot of avocado (ca *Phytophthora cinnamomi*)	Belisle et al. (2019)
		Heart rot of pineapple (ca *Phytophthora*)	Rohrbach and Schenk (1984)
Anilino Pyrimidines (AP's)	Cyprodinil	Powdery mildew of cucurbits (ca *Erysiphe*)	McGrath (2001)
Dicarboximides	Osmotic signal transduction pathway	Root rot of soybeans (ca *Rhizoctonia*)	Ajayi-Oyetunde and Bradley (2017)
		Early blight of tomato (ca *Alternaria*)	Jacobelis et al. (2023)
Cyanoacetamide oxime	Interferes with NA and amno acid symthesis	Late blight of tomato (ca *Phytophthora*)	Mendonca et al. (2015)
		Downy mildew of cucurbits (ca *Pseudoperonospora*)	Keinath and de Figueiredo Silva (2022)
Methyl Benzimidazole Carbamates	Tubulin polymerizaion	Sigatoka of banana (ca *Mycosphaerella*)	Marin et al. (2003)
		Sheath blight of rice (ca *Thanatephorus*)	Chin (1977)
		Powdery mildew of cucucubits (ca *Podosphaera*)	McGrath (2001)
N-Phenylcarbamates	Tubulin polymerizaion	Sigatoka of banana (ca *Mycosphaerella*)	Esguera et al. (2024)
Melanin Biosynthesis Inhibitors–reductase	Dehydratase in melanin biosynthesis	Blast of rice (ca *Magnaporthe*)	Ogoshi et al. (2018), Amoghavarsha et al. (2021)
Multisite organic	Multisite activity	Sigatoka of banana (ca *Mycosphaerella*)	Marin et al. (2003)

(continued)

Fungicide group (ex FRAC)	Site of action/ multisite	Diseases (ca, causal agent)	References
Examples of multisite products Mancozeb (EU banned since 2022), Chlorothalonil, Propineb, Thiram, Metiram		Powdery mildew of cucurbits (ca *Podosphaera*)	McGrath (2001)
		Sheath blight of rice (ca *Thanatephorus*)	Chin (1977)
Multisite inorganic			
Sulphur		Powdery mildew of cucurbits (ca *Podosphaera*)	McGrath (2001)
Plant defence inducers			
Chemical inducers	Salicylate related		
	Benzo-thiadiazole (BTH)	Red rot of sugar cane (ca *Colletotrichum falcatum*)	Ashwin et al. (2017)
	Probenazole	Blast of rice (ca *Magnaporthe oryzae*)	Watanabe (1977)
	Phosphonates (salts and esters of phosphorous acid)	Root diseases (ca *Pythium*)	Abbasi and Lazarovits (2006)
		Root rot of citrus (ca *Phytophthora*)	Chi et al. (2020)
		Heart rot of pineapple (ca *Phytophthora*	Rohrbach and Schenk (1984)
Microbial plant defence elicitors	Bacterial endophytes	*Phytopthora* Leaf fall of rubber (ca *Phytophthora*)	Abraham et al. (2013)
Biological control agents			
	Plant extracts	*Phytopthora* diseases of pepper and tomato	Giannakopoulou et al. (2014)
	Microbes or extracts from microbes	*Phytopthora* of pepper, tomato and tobacco	Volynchikova and Kim (2022)
	Trichoderma	Root rot of cowpeas (ca *Rhizoctonia*)	Wang and Zhuang (2018)
	Bacillus amyloliquefaciens (syn. *B. subtilis*)	Asian soybean rust (ca *Phakopsora*)	Dorighello et al. (2020)
	Trichoderma, Bacillus, Paenibacillus, Enterobacter, Streptomyces, Pseudomonas	downy mildew of cucurbits (ca *Pseudoperonospora*)	Sun et al. (2022)
	Metabolites or synthetic versions of metabolites (e.g. capsidiol)	*Phytopthora* diseases of pepper	Giannakopoulou et al. (2014)

(continued)

Fungicide group (ex FRAC)	Site of action/ multisite	Diseases (ca, causal agent)	References
IPM		*Phytophthora* diseases of durian	Guest et al. (2004)
		Asian soybean rust	Hershman et al. (2011)
		Huangloongbing of citrus	Alquézar et al. (2022)

Appendix 2

Overview of All Aspects of Disease Management

The availability of these control tools is labelled 'yes' or is described.

1. References refer to the columns host resistance to agronomy. References for control and biocontrol agents are provided in Appendix 1.
2. Host resistance: commercial nk (commercial availability not known or uncommon).
3. Cultural methods:
    - Prevention: clean planting material, removal of alternate hosts, quarantine procedures.
    - Sanitation or eradication of infected material/plants.
    - Agronomic: tillage, nutrition, irrigation, crop rotation.

Diseases (ca: causal agent; tv; transmitting vector)	Control agents§	Biocontrol	Host resistance (nk, not known)	Prevention	Sanitation or eradication	Tillage, agronomic, rotation	References
Anthracnose of mango and chili (ca *Colletotrichum*)	Yes		Commercial nk		Yes	Yes	Dofuor et al. (2023)
Asian rust of soybeans (ca *Phakopsora*)	Yes	Yes	Yes	Yes			Godoy (2016)
Basal stem rot of oil palm (ca *Ganoderma*)	Yes	Yes	Partial		Yes	Yes	Zakaria (2023)
Black shank of tobacco (ca *Phytophthora*)	Yes		Yes		Yes	Yes	Gallup et al. (2006)
Blast disease of rice (ca *Magnaporthe*)	Yes		Yes			Yes	Ning et al. (2020)
Blue mold of tobacco (ca *Peronospora*)	Yes		Commercial nk		Yes	Yes	Borras-Hildalgo et al. (2010)
Black pod of cocoa	Yes		Yes	Yes	Yes		Guest, 2007
Bremia tabaci vector of cassava mosaic virus disease	Yes	Yes	Commercial nk				Jacobson et al. (2018), Legg et al. (2015)
Brown spot of rice	Yes		Yes			Yes	Matsumoto et al. (2017)
Bacterial leaf blight of rice			Yes				Chukwu (2019)
Bacterial leaf streak of rice			Yes				Yesie et al. (2023)
Cassava mosaic and other virus diseases (tv *Bremia tabaci*)	Yes		Commercial nk	Yes	Yes		Jacobson et al. (2018), Legg et al. (2015)
Damping off and root rot of several crops (ca *Pythium*)	Yes				Yes		
Downy mildew of cucurbits	Yes	Yes	Yes				Call et al. (2012)
Early blight of tomato (ca *Alternaria*)	Yes		Yes				Foolad et al. (2008)
False smut of rice (ca *Ustilaginoidea*)	Yes		Commercial nk				Qiu et al. (2020)

(continued)

Diseases (ca: causal agent; tv; transmitting vector)	Control agents§	Biocontrol	Host resistance (nk, not known)	Prevention	Sanitation or eradication	Tillage, agronomic, rotation	References
Fusarium wilt of oil palm (ca *Fusarium*)			Yes	Yes	Yes		Flood (2006), Adusei-Fosu and Dickinson (2023)
Gummy stem blight of cucurbits (ca *Didymella*)	Yes		Yes		Yes	Yes	Seblani et al. (2023)
Heart rot of pineapple (ca *Phytophthora*)	Yes						Rattiab et al. (2018)
Huanglongbing of citrus (ca *Candidatus Liberibacter*)			nk	Yes	Yes	Yes	Alvarez et al. (2016)
Hwanglongbin of citrus (tv Asian citrus psyllid)	Yes	Yes	nk	Yes	Yes	Yes	Alquézar et al. (2022)
Late blight of tomato (ca *Phytophthora*)	Yes		Yes			Yes	Foolad et al. (2008)
Leaf spot of peanuts (ca *Cercospora*)	Yes		Commercial nk				Gonzales et al. (2023)
Lethal yellowing of coconut (ca *Candidatus Phytoplasma palmicola*)			nk	Yes	Yes		Dickinson (2015)
Narrow brown leaf spot of rice (ca *Cercospora*)	Yes		Commercial nk				Addison et al. (2021)
Patch canker and other *Phytophthora* diseases of durian	Yes		Yes		Yes	Yes	Misman et al. (2022), Krishnan et al. (2019)
Phytophthora diseases of pepper	Yes	Yes	Yes			Yes	Campbell et al. (2019)
Phytopthora diseases of rubber (ca *Phytophthora*)	Yes						Misman et al. (2022), Krishnan et al. (2019), Drenth and Guest (2013)
Powdery mildew of cucurbits (ca *Podosphaera*)	Yes		Yes				Nie et al. (2023)

Appendices

Disease	Col1	Col2	Col3	Col4	Col5	Col6	Reference
Red rot of sugar cane (ca *Colletotrichum falcatum*)	Yes			Yes	Yes	Yes	Viswanathan et al. (2017)
Root diseases (ca *Pythium*)	Yes					Yes	Belisle et al. (2019)
Root rot of avocado (ca *Phytopthora*)	Yes			Yes	Yes	Yes	
Root rot of cowpeas (ca *Rhizoctonia, Fusaium* and *Macrophomina*)	Yes	Yes	Commercial nk				Lamini et al. (2023)
Root rot of soybeans (ca *Rhizoctonia*)	Yes					Yes	Weaver and Rodriguez-Kabana (1992)
Root rot of soybenas (ca *Phytophthora, Fusarium*)			Yes			Yes	Zhi and Wang (2022)
Sheath bight of rice (ca *Thanatephorus*)	Yes				Yes	Yes	Chen et al. (2023)
Sigatoka of bananas (ca *Mycosphaerella*)[a]	Yes		Commercial uncommon		Yes	Yes	Soares et al. (2021)
Tungro virus of rice (RTSV and RTBV, tv *Nephotettiyes virescens*)	Yes		Yes				Hibino et al. (1991), Habibuddin et al. (1997), Hore et al. (2022)
Vascular streak dieback of cocoa (ca *Ceratobasidium*)	Yes		Yes	Yes	Yes		Guest and Keane (2018)

References

Abraham A, Philip S, Kuruvilla JC et al (2013) Novel bacterial endophytes from Hevea brasiliensis as biocontrol agent against Phytophthora leaf fall disease. BioControl 58:675–684. https://doi.org/10.1007/s10526-013-9516-0

Addison CK, Angira B, Cerioli T et al (2021) Identification and mapping of a novel resistance gene to the rice pathogen, Cercospora janseana. Theor Appl Genet 134:2221–2234. https://doi.org/10.1007/s00122-021-03821-2

Adusei-Fosu K, Dickinson M (2023) Detecting Fusarium oxysporum f. sp. elaeidis by using loop-mediated isothermal amplification. J Plant Pathol 105:1637–1643. https://doi.org/10.1007/s42161-023-01435-9

Ajayi-Oyetunde OO, Bradley CA (2017) Rhizoctonia solani: taxonomy, population biology and management of rhizoctonia seedling disease of soybean. Plant Pathol 67(1):3–17. https://doi.org/10.1111/ppa.12733

Alquézar B, Carmona L, Bennici S, Miranda MP, Bassanezi RB, Peña L (2022) Cultural management of Huanglongbing: current status and ongoing research. Phytopathology 112:11–25. https://doi.org/10.1094/PHYTO-08-21-0358-IA

Alvarez S, Rohrig E et al (2016) Citrus greening disease (Huanglongbing) in Florida: economic impact, management and the potential for biological control. Agric Res 5(2):109–118

Amoghavarsha C et al (2021) Chemicals for the management of paddy blast disease. In: Nayaka SC, Hosahatti R, Prakash G, Satyavathi CT, Sharma R (eds) Blast disease of cereal crops. Fungal biology. Springer. https://doi.org/10.1007/978-3-030-60585-8_5

Anderson JM, Pegg KG, Scott C et al (2012) Phosphonate applied as a pre-plant dip controls Phytophthora cinnamomi root and heart rot in susceptible pineapple hybrids. Australas Plant Pathol 41:59–68. https://doi.org/10.1007/s13313-011-0090-6

Ashwin NMR, Barnabas EL, Sundar RA et al (2017) Disease suppressive effects of resistance-inducing agents against red rot of sugarcane. Eur J Plant Pathology 149:285–297. https://doi.org/10.1007/s10658-017-1181-1

Batzer JC, Singh A, Rairdin A, Chiteri K, Mueller DS (2022) Mungbean: a preview of disease management challenges for an alternative U.S. cash crop. J Integr Pest Manag 13(1):4. https://doi.org/10.1093/jipm/pmab044

Belisle RJ, Hao W, McKee B, Arpaia M, Manosalva P, Adaskaveg JE (2019) New Oomycota fungicides with activity against Phytophthora cinnamomi and their potential use for managing avocado root rot in California. Plant Dis 103(8):2024–2032

Bellaire L, Fouré E, Abadie C, Carlier J (2010) Black leaf streak disease is challenging the banana industry. Fruits 65(06):327–342. https://doi.org/10.1051/fruits/2010034

Bittner RJ, Mila A (2017) Efficacy and timing of application of oxathiapiprolin against black shank of flue-cured tobacco. Crop Prot 93(1):9–18

Bittner RJ, Sweigard JA, Mila AL (2017) Assessing the resistance potential of Phytophthora nicotianae, the causal agent of black shank of tobacco, to oxathiopropalin with laboratory mutants. Crop Prot 102:63–71. https://doi.org/10.1016/j.cropro.2017.08.002

Borras-Hidalgo O, Thomma BPHJ, Yussuan S, Chacon O, Pujol M (2010) Tobacco blue mould disease caused by *Peronospora hyoscyam*i f. sp. *tabacina*. Mol Plant Pathol 11(1):13–18

Bowers JH, Bailey BA, Hebbar PK, Sanogo S, Lumsden RE (2001) The impact of plant diseases on world chocolate production. Plant Health Progress 2(1):12. https://doi.org/10.1094/PHP-2001-0709-01-RV

Brito F, Fraaije B, Robert M (2015) Sigatoka disease complex of Banana in Brazil: management practices and future directions. Outlooks Pest Manag 26(2):78–81

Bruns HA (2017) Southern corn leaf blight: a story worth retelling. Agron J 109:1–7

Bughio MA (2017) Cercospora leaf spot in mung beans. CABI International, Plantwise

Burkill IH (1966) A dictionary of the products of the Malay peninsula. Ministry of Agriculture and Cooperatives, Kuala Lumpur, p 2444

Call AD, Criswell AD, Todd C, Ando WH, Grumet R (2012) Resistance of cucumber Cultivarsto a new strain of cucurbit downy mildew. HortScience 47(2):171–178

Campbell TM, Ikley JT, Johnson WQG, Wise KA (2019) Impact of inoculum concentration on Goss's wilt development in corn and alternative hosts. Plant Health Progress 20:155–159. https://doi.org/10.1094/PHP-01-19-0002-R

Cartwright RD, Groth DE, Wamishe YA, Greer CA, Calvert LA, Cruz CM, Verdier V, Way MO (2018) Compendium of rice diseases and pests. American Phytopathological Society, St Paul, Minnesota, p 121

Chakraborty S, Newton AC (2011) Climate change, plant diseases and food security: an overview. Plant Pathol 60:2–14

Chakraborty S, Tiedemann AV, Teng PS (2000) Climate change: potential impact on plant diseases. Environ Pollut 108(2000):317–326

Chang S-T, Boswell J (1999) *Ganoderma lucidum* (Curt.: Fr.) P. Karst. (Aphyllophoromycetideae)—a mushrooming medicinal mushroom. Int J Med Mush 1:139–146

Chen J, Xuan Y, Yi J, Xiao G, Yuan P, Li D (2023) Progress in rice sheath blight resistance research. Front Plant Sci 14:1141697. https://doi.org/10.3389/fpls.2023.1141697

Chi NM, Thu PQ, Nam HB et al (2020) Management of Phytophthora palmivora disease in citrus reticulata with chemical fungicides. J Gen Plant Pathol 86:494–502. https://doi.org/10.1007/s10327-020-00953-z

Chin KM (1974) Chemical control of *Helminthosporium* leaf spot of rice. Malays Agric J 49(4):437–441

Chin KM (1975) Fungicidal control of the rice blast disease. Malays Agric J 50(2):221–228

Chin KM (1977) Chemical control of sheath blight disease of rice caused by *Thanatephorus cucumeris*. Malays Agric J 51(2):238–243

Chin KM (1986) A simple model of selection for fungicide resistance in plant pathogen populations. Phtopathology 77:666–669

Chin KM (2005) Rice disease datasheets. Crop protection compendium. CAB International, Wallingford

Chin KM, Ajimilah NH (1982) Rice variety mixtures in disease control. In: Proceedings of the international conference of plant protection in the tropics, Kuala Lumpur, pp 241–246

Chin KM, Holloman DW (2000) Fungicide resistance risk management. In: Proceedings BCPC symposium, pp 31–36

Chin KM, Supaad MA (1986) Diseases of rice in Malaysia. MARDI, Kuala Lumpur, p 89

Chin KM, Wolf MS (1984) The spread of *Erysiphe graminis* f.sp *hordei* in mixtures of barley varieties. Plant Pathol 33:89–100

Chin KM, Wolfe MS, Minchin P (1984) Host mediated interactions between pathogen genotypes. Plant Pathol 33:161–171

Chin K, Arroyo T, Forster B, Steden C (1996) Sensitivity of Mycosphaerella fijiensis to demethylation inhibitors in Central America: testing methodology and cross resistance. In: Proceedings of the XII Acorbat meeting, vol 27, Santo Domingo, pp 413–422

Chin KM, Wirz M, Laird D (2001) Sensitivity of *Mycosphaerella fijiensis* from Banana to trifloxystrobin. Plant Dis 85:1264–1270

Chukwu SC, Rafii MY, Ramlee SI et al (2019) Bacterial leaf blight resistance in rice: a review of conventional breeding to molecular approach. Mol Biol Rep 46:1519–1532. https://doi.org/10.1007/s11033-019-04584-2

Chung G (2011) Management of Ganoderma diseases in oil palm plantations. Planter 87:325–339

Claus A, Simões K, De Mio LLM (2022) SdhC-I86F mutation in Phakopsora pachyrhizi is stable and can be related to fitness penalties. Phytopathology 112(7):1413–1421. https://doi.org/10.1094/PHYTO-10-21-0419-R

Coffelt TA, Simpson CE (1997) Origin of the peanut. In: Kokalis-Burelle N, Porter DM, Rodriguez-Kabana R, Smith DH, Subrahmanyam P (eds) Compendium of peanut diseases. APS Press, St. Paul

Colussi J, Schnitkey G (2021a) New soybean record: historical growing of production in Brazil. Farmdoc daily 11:49

Colussi J, Schnitkey G (2021b) Brazil: Corn Production in Three Crops per Year. Farmdoc daily 11, 58, Department of Agricultural and Consumer Economics, University of Illinois at Urbana-Champaign

Corffelt TA, Simpson CE (1997) Compendium of peanut diseases. APS Press

Corley RHV, Tinker RB (2003) The oil palm, 4th edn. Blackwell Science

CPC Datasheets (2005) Crop protection compendium. © CAB International, Wallingford

Cruz CD, Magarey RD, Christie DN, Fowler GA, Fernandes JM, Bockus WW, Valent B, Stack JP (2016) Climate suitability for *Magnaporthe oryzae* Triticum pathotype in the United States. Plant Dis 100:1979–1987

Culbreath AK, Brenneman TB, Kemerait RC Jr (2002) Management of early leaf spot of peanut with pyraclostrobin as affected by rate and spray interval. Plant Health Progress 3. https://doi.org/10.1094/PHP-2002-1018-01-RS

Culbreath AK, Kemerait RC Jr, Brenneman TB (2008) Management of leaf spot diseases of peanut with prothioconazole applied alone or in combination with tebuconazole or trifloxystrobin. Peanut Sci 35(2):149–158. https://doi.org/10.3146/PS08-005.1

Davenport TL (2007) Reproductive physiology of mango. Braz J Plant Physiol 19(4):363–376

De Langhe E, Vrydaghs L, de Maret P, Perrier X, Denham T (2009) Why bananas matter: an introduction to the history of banana domestication. Ethnobot Res Appl 7:165–177

De Los Reyes A (2024) let me help you tell a story: strategies for improving writing clarity. The Journal of Clinical Child and Adolescent Psychology Future Directions Forum. American Psychological Association. https://www.apa.org/career-development/writing-clarity.pdf

De Souza JT, Monteiro FP, Ferreira MA, Gramacho KP, Martins ED, Luz N (2018) Cocoa diseases: witches' broom. In: Umaharan P (ed) Achieving sustainable cultivation of cocoa. Burleigh Dodds Science Publishing, Cambridge

Dela Cueva FM, de Castro AM, de Torres RD (2020) *Peronosclerospora philippinensis* (Philippine downy mildew of maize). Invasive species compendium. CABI, Wallingford. https://doi.org/10.1079/ISC.44646.20210200696

Dickinson M (2015) Loop-mediated isothermal amplification (LAMP) for detection of Phytoplasmas in the field. In: Lacomme C (ed) Plant pathology. Methods in molecular biology, vol 1302. Humana Press, New York, p 99. https://doi.org/10.1007/978-1-4939-2620-6_8

Dofuor AK, Quartey NK, Osabutey AF et al (2023) Mango anthracnose disease: the current situation and direction for future research. Front Microbiol 14:1168203. https://doi.org/10.3389/fmicb.2023.1168203

Dorighello DV, Forner C, de Campos LRMVB et al (2020) Management of Asian soybean rust with Bacillus subtilis in sequential and alternating fungicide applications. Australas Plant Pathol 49:79–86. https://doi.org/10.1007/s13313-019-00677-5

Dos Santos GC, Lima Horn LM, Trezzi Casa R, Soardi K, Lopes MA, Nascimento SCD, Santi VM et al (2024) First report of seed decay caused by Diaporthe ueckeri on soybean in Brazil. Plant Dis. https://doi.org/10.1094/PDIS-04-24-0814-PDN

Drenth A, Guest DI (2013) Phytophthora palmivora in tropical tree crops. In: Lamour K (ed) Phytophthora: a global perspective. Centre for Agriculture and Bioscience International (CABI), pp 187–196. https://doi.org/10.1079/9781780640938.0187

Drenth A, Guest DI (2016) Fungal and oomycete diseases of tropical tree fruit crops. Annu Rev Phytopathol 54:373–395

EASAC (2022) Regenerative agriculture in Europe, EASAC policy report, p 58

Ebert AW (2017) Vegetable production, diseases, and climate change. In: World agricultural resources and food security: international food security. Emerald Publishing Limited, pp 103–124

EEA (2021) Growing degree days. https://www.eea.europa.eu/data-and-maps/figures/growing-degree-days

Erwin DC, Ribeiro OK (1996) *Phytophthora* diseases worldwide. The American Phytopathological Society, St Paul, p 562

Esguera JG, Balendres MA, Paguntalan DP (2024) Overview of the Sigatoka leaf spot complex in banana and its current management. Tropical Plants 3:e002. https://doi.org/10.48130/tp-0024-0001

Fanigliulo A, Sacchetti M (2009) Mandipropamid: new fungicide against *Phytophthora infestans* on tomato. Acta Hortic 808:355–358. https://doi.org/10.17660/ActaHortic.2009.808.58

FAO (2022) Major tropical fruits – statistical compendium 2021. Rome https://www.fao.org/markets-and-trade/publications/detail/en/c/1608288/

FAO (2023a) Durian global trade overview. Rome

FAO (2023b) Integrated pest management. https://www.fao.org/agriculture/crops/thematic-sitemap/theme/spi/scpi-home/managing-ecosystems/integrated-pest-management/it/

FAO, IFAD, UNICEF, WFP and WHO (2021) The State of Food Security and Nutrition in the World (SOFI). Transforming food systems for food security, improved nutrition and affordable healthy diets for all. Rome, FAO. https://doi.org/10.4060/cb4474en

Flood J (2006) A review of Fusarium wilt of oil palm caused by Fusarium oxysporum f. sp. elaeidis. Phytopathology 96:660–662

Foolad MR, Merk HL, Ashrafi H (2008) Genetics, genomics and breeding of late blight and early blight resistance in tomato. Crit Rev Plant Sci 27(2):75–107. https://doi.org/10.1080/07352680802147353

FRAC (2024) Fungicide resistance action committee. https://www.frac.info/

Gaba S, Rebouda X, Fried G (2016) Agroecology and conservation of weed diversity in agricultural lands. Bot Lett 163(4):351–354. https://doi.org/10.1080/23818107.2016.1236290

Gallup CA, Sullivan MJ, Shew HD (2006) Black Shank of Tobacco. The Plant Health Instructor. https://doi.org/10.1094/PHI-I-2006-0717-01

Ganry J, Fouré E, de Lapeyre de Bellairaire L, Lescot T (2012) An integrated approach to control the black leaf streak disease (BLSD) of bananas, while reducing fungicide use and environmental impact. In: Fungicides for plant and animal diseases, pp 194–226. https://doi.org/10.5772/29794

Gao S, Xu L, Zeng R, Gao P, Song Z, Dai F (2022) Baseline sensitivity of Rhizoctonia solani to four DMI fungicides. J Basic Microbiol 67(6):701–710. https://doi.org/10.1002/jobm.202100642

Ghini R, Bettiol W, Hamada E (2011) Diseases in tropical and plantation crops as affected by climate changes: current knowledge and perspectives. Plant Pathol 60(1):122–132

Giannakopoulou A, Schornack S et al (2014) Variation in Capsidiol sensitivity between Phytophthora infestans and Phytophthora capsici is consistent with their host range. PLoS One 9(9):1–11. https://doi.org/10.1371/journal.pone.0107462

Gilbertson RL, Macedo MA, Maliano MR, Rojas MR (2021) Weed-infecting viruses in a tropical agroecosystem presents different threats to crops and evolutionary histories. PLoS One 16(4):e0250066. https://doi.org/10.1371/journal.pone.0250066

Giller KE, Hijbeek R, Andersson JA, Sumberg J (2021) Regenerative Agriculture: an agronomic perspective. Outlook Agric 32(1):13–25

Gisi U, Sierotzki H (2015) Oomycete fungicides: Phenylamides, Quinone outside inhibitors, and carboxylic acid amides. In: Ishii H, Hollomon D (eds) Fungicide resistance in plant pathogens. Springer, Tokyo

Godding D, Stutt ROJH, Alicai T, Abidrabo P et al. (2023) Developing a predictive model for an emerging epidemic on cassava in sub-Saharan Africa. http://www.nature.com/scientificreports

Godoy CV, Seixas CDS, Soares RM et al (2016) Asian soybean rust in Brazil: past, present, and future. Pesq Agrop Brasileira 51(5):407–421. https://doi.org/10.1590/S0100-204X2016000500002

Gonzales M, Kemerait R Jr et al (2023) Strong resistance to early and late leaf spot in Peanut-compatible wild-derived induced Allotetraploids. Plant Dis 107(2):335–343. https://doi.org/10.1094/PDIS-03-22-0721-RE

Guest D (2007) Black pod: diverse pathogens with a global impact on cocoa yield. Phytopathology 97:1650–1653

Guest D, Keane P (2018) Cocoa diseases: vascular-streak dieback. In: Achieving sustainable cultivation of cocoa, pp 287–301. https://doi.org/10.19103/AS.2017.0021.18

Guest DI, Minh CN, Sangchote S, Vawdrey L, Diczbalis Y (2004) Integrated management of phytophthora diseases of durian: recommendations and benefit-cost analysis. In: Diversity and management of Phytophthora in Southeast Asia, pp 222–226

Guicherit E et al (2014) Solatenol/benzovindiflupyr – the second generation benzonorbornene SDHI carboxamide with outstanding performance against key crop diseases. Rheinhardsbrunn

Gurr GM, Johnson AC, Ash GJ, Wilson BAL, Ero MM, Pilotti CA, Dewhurst CF, You MS (2016) Coconut lethal yellowing diseases: a *Phytoplasma* threat to palms of global economic and social significance. Front Plant Sci 7:1521. https://doi.org/10.3389/fpls.2016.01521

Habibuddin H (1978) Incidence of rice ragged stunt disease of rice in Malaysia. MARDI Research Bulletin 6(1):113–117

Habibuddin H, Mahir AM, Ahmad IB et al (1997) Genetic analysis of resistance to rice tungro spherical virus in several rice varieties. J Trop Agric Food Sci 25(1):1–7

Hanafiah NM, Mispan MS, Lim PE, Baisakh N, Cheng A (2020) The 21st century agriculture: when rice research draws attention to climate variability and how weedy rice and underutilized grains come in Handy. Plan Theory 9:365. https://doi.org/10.3390/plants9030365

Harlan JR (1976) The plants and animals that nourish man. Sci Am 235(3):88–97

Hartman GL, Rupe JC, Sikora EJ, Domier LL, Davies JA, Steffey KL (2016) Compendium of soya bean diseases and pests. American Phytopathological Society, St Paul

He H, Zhai Q, Tang Y, Gu X, Pan H, Zhang H (2023) Effective biocontrol of soybean root rot by a novel bacterial strain Bacillus siamensis HT1. Physiol Mol Plant Pathol 125:101984. https://doi.org/10.1016/j.pmpp.2023.101984

Hershman DE, Sikora EJ, Giesler LJ (2011) Soybean rust PIPE: past, present, and future. J Integr Pest Manag 2(2):D1–D7. https://doi.org/10.1603/IPM11001

Heslop-Harrison JS, Schwarzacher T (2007) Domestication, genomics and the future for Banana. Ann Bot 100(5):1073–1084

Hibino H, Daquioag RD, Mesina EM, Aguiero VM (1991) Resistances in rice to tungro-associated viruses. Plant Dis 74(11):923–926

Hore TK, Inabangan-asilo MA, Wulandari R et al (2022) Introgression of tsv1 improves tungro disease resistance of a rice variety BRRI dhan71. Sci Rep 12:18820. https://doi.org/10.1038/s41598-022-23413-4

Horgan FG, Peñalver-Cruz A (2022) Compatibility of insecticides with rice resistance to Planthoppers as influenced by the timing and frequency of applications. Insects 13(2):106. https://doi.org/10.3390/insects13020106

How TC (2002) Classification of banana varieties in Malaysia using retrotransposons. MSc thesis. Universiti Putra Malaysia

Husin N, Sapak Z (2022) Bacillus cereus for controlling bacterial heart rot in pineapple var. MD2. Trop Life Sci Res 33(1):77–89. https://doi.org/10.21315/tlsr2022.33.1.5

Institute of Food Technologists (2019) The most important nut is a legume. Food Technology Magazine 73(10)

IRRI (2007) A hybrid history. Rice Today 6(4). International Rice Research Institute DAPO Box 7777, Metro Manila, Philippines Web: ricetoday.irri.org

IRRI (2016) Rice that changed the world: celebrating 50 years of IR8. Rice Today, International Rice Research Institute DAPO Box 7777, Metro Manila, Philippines Web: ricetoday.irri.org

IRRI (2024) Rice knowledge bank. Diseases. http://www.knowledgebank.irri.org/step-by-step-production/growth/pests-and-diseases/diseases

Jackson SA (2016) Rice: the first crop genome. Rice 9:14. https://doi.org/10.1186/s12284-016-0087-4

Jacobelis W Jr, Aires ES, Ferraz AKL, Marques ICS, Freitas FGBF Jr, Silva DMR, Ono EO, Rodrigues JD (2023) Application of Strobilurins and Carboxamides improves the physiology and productivity of tomato plants in a protected environment. Horticulturae 9(2):141. https://doi.org/10.3390/horticulturae9020141

Jacobson L, Duffy S, Sseuragi P (2018) Whitefly transmitted viruses threatening cassava production in Africa. Curr Opin Virol 33:167–176

Jansz ER, Uluwaduge DI (1997) Biochemical aspects of cassava (Manihot esculenta) with special emphasis on cyanogenic glucosides. J Natn Sci Council, Sro Lanka 25(1):1–24

Jeger MJ, Fereres A, Malmstrom CE, Mauck KE, Wintermantel WM (2023) Epidemiology and management of plant viruses under a changing climate. Phytopathology 113:1620. https://doi.org/10.1094/PHYTO-07-23-0262-V

Johnson R (1983) Genetic background of durable resistance. In: Lamberti F, Waller JM, Van der Graaff NA (eds) Durable resistance in crops. NATO advanced science institutes series, vol 55. Springer, Boston. https://doi.org/10.1007/978-1-4615-9305-8_2

Johnston M, Foley JA, Holloway T, Kucharik CJ, Monfreda C (2009) Resetting global expectations from biofuels. Environ Res Lett 4:014004. https://doi.org/10.1088/1748-9326/4/1/014004

Jones JB, Zitter TA, Momol TM, Miller SA (2016) Compendium of tomato diseases and pests. American Phytopathological Society, St Paul

Keane PJ (1981) Epidemiology of vascular-streak dieback of cocoa. Ann Appl Biol 98(2):227–241

Keinath AP, de Figueiredo Silva F (2022) Economic impacts of reduced fungicide efficacy against downy mildew on slicing cucumber. Crop Prot 155:105934. https://doi.org/10.1016/j.cropro.2022.105934

Keinrath AP, Wintermantel WM, Zitter TA (2017) Compendium of cucurbit diseases and pests. American Phytopathological Society, St Paul, Minnesota

Kessmann H, Staub T, Hofmann C, Maetzke T, Herzog J, Ward E, Uknes S, Ryals J (1994) Induction of systemic acquired disease resistance in plants by chemicals. Annu Rev Phytopathol 32:439–459

Koike S, Subbarao K, Davis RM, Turini T (2003) Vegetable diseases caused by soilborne pathogens. UCANR Publications

Kongtragoul P, Ishikawa K, Ishii H (2021) Metalaxyl resistance of Phytophthora palmivora causing durian diseases in Thailand. Horticulturae 7(10):375. https://doi.org/10.3390/horticulturae7100375

Kranz J, Schmutterer H, Koch W (1977) Diseases pests and weeds in tropical crops. Verlag Paul Parey, Berlin, p 666

Krishnan A et al (2019) An insight into Hevea – Phytophthora interaction: the story of Hevea defense and Phytophthora counter defense mediated through molecular signalling. Curr Plant Biol 17:33–41. https://doi.org/10.1016/j.cpb.2018.11.009

Kumar R, Srinivas K, Sivaramane N (2013) Assessment of the maize situation, outlook and investment opportunities in India. Country report – regional assessment Asia (MAIZE-CRP), National Academy of Agricultural Research Management, Hyderabad, India

Kumar R, Srinivas K, Boiroju NK, Gedam PC (2014a) Production performance of maize in India: approaching an inflection point. Int J Agric Statist Sci 10(1):241–248

Kumar R, Srinivas K, Monayem Miah MA, Shah H (2014b) Assessment of the maize situation, outlook and opportunities in Asia. In: Conference: 12th Asian maize conference and expert consultation on maize for food, feed, nutrition and environmental security, Bangkok

Lamini S, Kusi F, Cornelius EW et al (2023) Identification of sources of resistance in cowpea lines to Macrophomina root rot disease in northern Ghana. Heliyon 8(2022):e12217. https://doi.org/10.1016/j.heliyon.2022.e12217

Leadbeater AJ (2014) Plant health management: fungicides and antibiotics. In: Encyclopedia of agriculture and food systems

Legg P, Lava Kuma P, Makeshkumar T, Trpathi L, Fergusaon M, Kanju E, Ntawuruhunga CW (2015) Cassava virus diseases. Adv Virus Res 19:85–142

Li Z, Zhou JL, Lin Z (2019) Development and innovation of Ganoderma industry and products in China. Adv Exp Med Biol 1181:187–204

Liberei R (2007) South American leaf blight of the rubber tree (*Hevea* spp.): new steps in plant domestication using physiological features and molecular markers. Ann Bot 100(6):1125–1142. https://doi.org/10.1093/aob/mcm133

Liu Z, Zhu Y, Shi H, Qiu J, Ding X, Kou Y (2021) Recent progress in rice broad-spectrum disease resistance. Int J Mol Sci 22(21):11658. https://doi.org/10.3390/ijms222111658

Locosselli GM, Brienen RJW, Leite MS, Gloor M, Krottenthaler S, Oliveira AA, Barichivich J, Anhuf D, Ceccantini G, Schöngart J, Buckeridge M (2020) Global tree-ring analysis reveals rapid decrease in tropical tree longevity with temperature. Proc Natl Acad Sci USA 117(52):33358–33364. https://doi.org/10.1073/pnas.2003873117

Longstaff PH (2011) Resilience in systems. dealing with the new normal. Innovation brown bag seminars, knowledge management, ETH University of Zurich

Malvick D, Syverson R, Mollov D, Ishimaru CA (2010) Goss's bacterial blight and wilt of corn caused by *Clavibacter michiganensis* subsp. *nebraskensis* occurs in Minnesota. Plant Dis 94(8):1064. https://doi.org/10.1094/PDIS-94-8-1064A

Marín DH, Romero RA, Guzmán M, Sutton TB (2003) Black Sigatoka an increasing threat to banana cultivation. Plant Dis 87(3):208–222

Martin J (2008) Target Earth: the grand scale problems of the 21st Century, Public Lectures of the Oxford Martin School. University of Oxford. https://www.oxfordmartin.ox.ac.uk/events/public-lecture-target-earth-by-james-martin

Matsumoto K, Ota Y, Seta S, Nakayama Y, Ohno T, Mizobuchi R, Sato H (2017) Identification of QTLs for rice brown spot resistance in backcross inbred lines derived from a cross between Koshihikari and CH45. Breed Sci 67(5):540–543. https://doi.org/10.1270/jsbbs.17057

May RM (1972) Limit cycles in predator prey communities. Science 177:900–902

McGrath MT (2001) Fungicide resistance in cucurbit powdery mildew. Plant Dis 85(3):236–245

McGrath MT (2007) Managing cucurbit powdery mildew and fungicide resistance. Acta Hortic 731:211–216. https://doi.org/10.17660/ActaHortic.2007.731.29

McGrath MT (2011) Chapter 16: challenge of fungicide resistance in managing vegetable diseases in United States and anti-resistance strategies. In: Thind TS (ed) Fungicide resistance in crop protection: threat and management. CABI International, pp 191–207

McGrath MT, Fox GM (2010) Efficacy of fungicides for managing cucurbit powdery mildew and treatment impact on pathogen sensitivity to fungicides, 2009. Plant Dis Manag Rep 4:V147

Mena E, Stewart S, Montesano M, de Leon IP (2023) Plant Pathol 73(1):31–46. https://doi.org/10.1111/ppa.13803

Mendoca LBP, Coelho L, Stratcieri J, Ferreira Junior JB, Tebaldi ND (2015) Chemical control of Phytophthora wilt in tomatoes. Biosci J, Uberlândia 3(4):1015–1023

Miao J, Dong X, Lin D, Wang Q, Liu P, Chen F (2016) Activity of the novel fungicide oxathiapiprolin against plant-pathogenic oomycetes. Pest Manag Sci 72(8):1572–1577

Milne AE, Teiken C, Deledalle F, van den Bosch F, Gottwald TR, McRoberts N (2018) Growers' risk perception and trust in control options for huanglongbing citrus-disease in Florida and California. Crop Prot 114:177–186

Misman N, Samsulrizal NH, Noh AL, Wahab MA, Ahmad K, Ahmad Azmi NS (2022) Host range and control strategies of *Phytophthora palmivora* in Southeast Asia perennial crops. Pertanika J Trop Agric Sci 45(4):991–1019

Moens M, Perry RN, Starr JL (2009) Meloidogyne species: a diverse group of novel and important plant parasites. In: Perry RN, Moens M, Starr JL (eds) Root-knot nematodes. CABI Publishing, Wallingford, pp 1–17

Mollick E (2024) Co-intelligence: living and working with AI, Penguin, p 256

Mpunami AA, Tymon A, Jones P, Dickinson M (1999) Genetic diversity in the coconut lethal yellowing disease phytoplasmas of East Africa. Plant Pathol 48(1):109–114

Mukherjee SK (1972) Origin of mango. Econ Bot 26(3):260–264

Müller U, Gisi U (2011) Newest aspects of nucleic acid synthesis inhibitors: Metalaxyl-M. (eds) Dr. Wolfgang Krämer, Dr. Ulrich Schirmer, Dr. Peter Jeschke, Dr. Matthias Witschel. https://doi.org/10.1002/9783527644179.ch

Muniappan R, Heinrichs EA (eds) (2016) Integrated pest management of tropical vegetable crops. Springer

Munkvold GP, White DG (2016) Compendium of corn diseases. American Phytopathological Society, St Paul

Murphy DJ (2014) The future of oil palm as a major global crop: opportunities and challenges. J Oil Palm Res 26(1):1–24

Murray GM (2009) Threat specific contingency plan: Philippine downy mildew of maize (*Peronosclerospora philippinensis*) and downy mildew of sorghum (*P. sorghi*). Australia: Plant Health Australia.https://www.planthealthaustralia.com.au/wp-content/uploads/2013/03/Downy-mildew-of-maize-and-sorghum-CP-2009.pdf

Nature (2005) International Rice genome sequencing project and Sasaki T. The map-based sequence of the rice genome. Nature 436:793–800. https://doi.org/10.1038/nature03895

Newton P, Civita N, Frankel-Goldwater L, Bartel K, Colleen JC (2020) What is regenerative agriculture? Front Sustain Food Syst 4:1–11

Nie J, Yuan Q, Zhang W, Pan J (2023) Genetics, resistance mechanism, and breeding of powdery mildew resistance in cucumbers (Cucumis sativus L.). Hortic Plant J 9(4):603–615

Niessen L (2015) Current state and future perspectives of loop-mediated isothermal amplification (LAMP)-based diagnosis of filamentous fungi and yeasts. Appl Microbiol Biotechnol 99(2):553–574. https://doi.org/10.1007/s00253-014-6196-3

Ning X, Yunyu W, Aihong L (2020) Strategy for use of rice blast resistance genes in rice molecular breeding. Rice Sci 27(4):263–277

Noor Camillia NA, Salma I, Mohd Norfaizal G (2019) Development of SCAR markers for rapid identification of resistance to Phytophthora in durian using inter simple sequence repeat markers. Asian J Adv Basic Sci 7(1):30–34

Nordmeyer D, Halim B, Liew CY, Chin KM, Ruess W (1994) Difenoconazole – a new systemic broad-spectrum fungicide for a wide range of crops in Malaysia. In: Proceedings 4th international conference on plant protection in the tropics, Kula Lumpur, pp 263–265

Oberč BP, Schnell A (2020) Approaches to sustainable agriculture. Exploring the pathways towards the future of farming. IUCN EURO, Brussels

Ogden LE (2020) Biocontrol 2.0: a shifting risk–benefit balance. Bioscience 70:17–22. https://doi.org/10.1093/biosci/biz135

Ogoshi C, Carlos FS, Ulguim AR, Zanon AJ, Bittencourt CRC, Almeida RD (2018) Effectiveness of fungicides for rice blast control in lowland rice crop in Brazil. Trop Subtrop Agroecosyst 21:505–511

Ojiambo PS, Paul PA, Holmes GJ (2010) A quantitative review of fungicide efficacy for managing downy mildew in cucurbits. Phytopathology 100:1066–1076

Otang NV, Tripathi JN, Kariuki SM, Tripathi L (2023) Cassava molecular genetics and genomics for enhanced resistance to diseases and pests. Mol Plant Pathol 25(1)

Ou SH (1985) Rice diseases, 2nd edn. Commonwealth Mycological Institute, Kew, p 380

Padmanabhan SY (1973) The great Bengal famine. Ann Rev Phytopathol 11:11–26

PANS Manual No.1 (1971) Pest control in bananas. Overseas Development Administration, London, p 128

Parada-Rojas CH, Quesada-Ocampo LM (2019) Characterizing sources of resistance to Phytophthora blight of pepper caused by Phytophthora capsici in North Carolina. Plant Health Progress 20:112–119. https://doi.org/10.1094/PHP-09-18-0054-RS

Pascal B (1670) Pensées, 1670. Krailsheimer, AJ (Translator, Introduction). Penguin Classics 1995

Patterson RRM (2006a) Ganoderma boninense (basal stem rot of oil palm). Cabi Compendium, Datasheet. https://doi.org/10.1079/cabicompendium.24924

Patterson RRM (2006b) Ganoderma – a therapeutic fungal biofactory. Phytochemistry 67(18):1985–2001. https://doi.org/10.1016/j.phytochem.2006.07.004

Patterson RRM (2019) Ganoderma boninense disease of oil palm to significantly reduce production after 2050 in Sumatra if projected climate change occurs. Microorganisms 7:24. https://doi.org/10.3390/microorganisms7010024

Pavlou GC, Vakalounakis DJ, Ligoxigakis EK (2002) Control of root and stem rot of cucumber, caused by Fusarium oxysporum f. sp. radicis-cucumerinum, by grafting onto resistant rootstocks. Plant Dis 86:379–382

Pérez L, Hernández A, Hernández L, Pérez M (2002) Effect of trifloxystrobin and azoxystrobin on the control of black Sigatoka (Mycosphaerella fijiensis Morelet) on banana and plantain. Crop Prot 21(1):17–23. https://doi.org/10.1016/S0261-2194(01)00055-2

Peritore-Galve FC, Tancos MA, Smart CD (2021) Bacterial canker of tomato: revisiting a global and economically damaging Seedborne pathogen. Plant Dis 105(6):1581–1595

Ploetz RC (ed) (2003) Diseases of tropical fruit crops. Cabi Publishing, p 481

Ploetz RC (2005) Panama disease, an old nemesis rears its ugly head: part 1, the beginnings of the banana export trade. Plant Health Progress 6. https://doi.org/10.1094/PHP-2005-1221-01-RV

Ploetz RC (2006) Panama disease, an old nemesis rears its ugly head: part 2, the cavendish era and beyond. Plant Health Progress. https://doi.org/10.1094/PHP-2006-0308-01-RV

Puckridge DW, Kupkanchanul T, Palaklang W, Kupkanchanaku K (2000) Production of rice and associated crops in deeply flooded areas of the Chao Phraya delta. In: Proceedings of a conference on deep-water Rice and associated crops, Bangkok

Qassim WS, Mohamed AH, Hamdoon ZK (2023) Biological control of root rot fungi in cowpea. SABRAO J Breed Genet 56(1):302–309. https://doi.org/10.54910/sabrao2024.56.1.27

Qiu J, Lu F, Wang H et al (2020) A candidate gene for the determination of rice resistant to rice false smut. Mol Breeding 40:105. https://doi.org/10.1007/s11032-020-01186-w

Rao BS (1975) Maladies of *Hevea* in Malaysia. Rubber Research Institute of Malaysia, p 108

Rattiab MF, Ascunceab MS, Landivarcand JJ, Gossab EM (2018) Pineapple heart rot isolates from Ecuador reveal a newgenotype of Phytophthora nicotianae. Plant Pathol 67:1803–1813. https://doi.org/10.1111/ppa.12885

Reglinski T, Havis N, Rees HJ, de Jong H (2023) The practical role of induced resistance for crop protection. Phytopathology 113:719. https://doi.org/10.1094/PHYTO-10-22-0400-IA

Renfro BL, Ullstrup AJ (1976) A comparison of maize diseases in temperate and in tropical environments. PANS 22(4):491–498

Rohrbach KG, Schenck S (1984) Control of pineapple heart rot, caused by Phytophthora parasitica and P. Cinnamomi, with Metalaxyl, Fosetyl Al, and phosphorous acid. Plant Dis 69:320–323. https://doi.org/10.1094/PD-69-320

Ruf F, Zadi H (1998) Cocoa: from deforestation to reforestation. In: First international workshop on sustainable cocoa growing. Smithsonian Tropical Research Institute, Panama City

Ruf F, Konan G, Ardhy W (1995) Forest rent, replanting and regulation of cocoa supply. In: Conference paper, 8th meeting of the advisory group on the world cocoa economy, Yaounde

Sánchez DG, Mendieta EH, Granados CR, Pérez RH (2017) Comparison of two fungicide in the control of black Sigatoka (Mycosphaerella fijiensis Morelet), in banana cv.' Dwarf giant' (AAA), Teapa, Tabasco, México. Global Adv Res J Agric Sci 6(8):233–239

Sani I, Ismail SI, Abdullah S, Jalinas J, Jamian S, Saad N (2020) A review of the biology and control of whitefly, Bemisia tabaci (Hemiptera: Aleyrodidae), with special reference to biological control using Entomopathogenic fungi. Insects 11(9):619. https://doi.org/10.3390/insects11090619

Santini A, Liebhold A, Migliorini D (2018) Tracing the role of human civilization in the globalization of plant pathogens. ISME J 12:647–652. https://doi.org/10.1038/s41396-017-0013-9

Sanya DRA, Syed-Ab-Rahman SF, Jia A et al (2022) A review of approaches to control bacterial leaf blight in rice. World J Microbiol Biotechnol 38:113. https://doi.org/10.1007/s11274-022-03298-1

Sapak Z, Nusaibah SA (2024) Common diseases in pineapple and their management. In: Wong MY (ed) Advances in tropical crop protection. Springer, Cham. https://doi.org/10.1007/978-3-031-59268-3_7

Saponari M, Giampetruzzi A, Loconsole G, Boscia D, Saldarelli P (2019) Xylella fastidiosa in olive in Apulia: where we stand. Phytopathology 109:175–186. https://doi.org/10.1094/PHYTO-08-18-0319-FI

Sarkar MSK, Begum RA, Pereira JJ (2020) 2020 impacts of climate change on oil palm production in Malaysia. Environ Sci Pollut Res 27:9760–9770. https://doi.org/10.1007/s11356-020-07601-1

Schneider K, van der Werf W, Cendoya M, Mourits M, Navas-Cortés JA, Vicent A, Lansink AO (2020) Impact of *Xylella fastidiosa* subspecies *pauca* in European olives. PNAS 117(17):9250–9259. https://doi.org/10.1073/pnas.1912206117

Schumann GL, D'Arcy CJ (2017) Hungry planet: stories of plant diseases. APS Publication

Schumann GL, Leonard KJ (2001) Stem rust of wheat (black rust). Plant Health Instructor. https://doi.org/10.1094/9780890544907

Schwartz HF, Mohan SK (2016) Compendium of onion and garlic diseases and pests. American Phytopathological Society, St Paul

Seblani R, Keinath AP, Munkvold G (2023) Gummy stem blight: one disease, three pathogens. Mol Plant Pathol 24(8):825–837. https://doi.org/10.1111/mpp.13339

Seid A, Fininsa C, Mekete T, Decraemer W, Wesemael WML (2015) Tomato (Solanum lycopersicum) and root-knot nematodes (Meloidogyne spp.)– a century-old battle. Nematology 17:995–1009

Shew B (2020) NC state extension publications. Peanut leaf spot. https://content.ces.ncsu.edu/early-leaf-spot-of-peanut-1

Sierotsky H, Scalliet G (2013) A review of current knowledge of resistance aspects for the next-generation succinate dehydrogenase inhibitor fungicides. Phytopathology 103(9):880–887

Silva TC, Moreira SI, de Souza DM, Christiano FS Jr, Gasparoto MCG, Fraaije BA, Goldman GH, Ceresini PC (2024) Resistance to site-specific succinate dehydrogenase inhibitor fungicides is pervasive in populations of black and yellow Sigatoka pathogens in Banana plantations from southeastern Brazil. Agronomy 14(4):666. https://doi.org/10.3390/agronomy14040666

Simmonds NW (1966) Bananas, 2nd edn. Longmans, London, p 512

Singh A, Chow C, Nathaniel K, Lip Vun Y, Javad S, Jabeen K (2025) Management of *Phytophthora* and *Phytopythium* oomycete diseases in durian (*Durio zibethinus*). Crop Prot 190:107086. https://doi.org/10.1016/j.cropro.2024.107086

Snowden DJ (2003) Narrative patterns: the perils and possibilities of using story in organisations. In: Lesser E, Prusak L (eds) Creating value with knowledge. Oxford University Press, p 14

Snowden DJ (2012) Agile: good practice, poor theory. Lecture to Auckland, Auckland

Snowden DJ, Boone ME (2007) A leader's framework for decision making. Harv Bus Rev 85(11):68–76

Snowden DJ, Goh Z (2020) Cynefin – weaving sense-making into the fabric of our world. In: Riva Greenberg, Boudewijn Bertsch (eds), The Cynefin Co, p 376

Soares JMS, Rocha AJ, Nascimento FS, Santos AS, Miller RNG, Ferreira CF, Haddad F, Amorim VBO, Amorim EP (2021) Genetic improvement for resistance to black Sigatoka in bananas: a systematic review. Front Plant Sci 21(12):657916. https://doi.org/10.3389/fpls.2021.657916

Soh WK (2017) Taxonomy of Syzygium. In: Nair KN (ed) The genus Syzygium: Syzygium cumini and other underutilized species. CRC Press, New York

Song J-H, Wang Y-F, Yin W-X, Huang J-B, Luo C-X (2021) Effect of chemical seed treatment on rice false smut control in field. Plant Dis 105:3218–3233. https://doi.org/10.1094/PDIS-11-19-2411-RE

Srinivasan R (2016) Integrated pest management in tropical vegetable crops. In: Integrated pest management in the tropics, pp 219–247

Sun Z, Yu S, Hu Y, Wen Y (2022) Biological control of the cucumber downy mildew pathogen Pseudoperonospora cubensis. Horticulturae 8(5):410. https://doi.org/10.3390/horticulturae8050410

Talhinhas P, Baptista D, Innes DINI, Vieira A, Silva DN, Loureiro A et al (2017) The coffee leaf rust pathogen *Hemileia vastatrix*: one and a half centuries around the tropics. Mol Plant Pathol 18(8):1039–1051

Tatineni S, Hein GL (2022) Plant viruses of agricultural importance: current and future perspectives. Phytopathology 113:117–141. https://doi.org/10.1094/PHYTO-05-22-0167-RVW

Theerthagiri A, Angannan C, Sasthamoorthy Pillai K, Govindasamy S, Thiruvengadam R, Ramasamy S (2008) Effectiveness of azoxystrobin in the control of Erysiphe cichoracearum and Pseudoperonospora cubensis on cucumber. J Plant Protect Res 48(2):147–159. https://doi.org/10.2478/v10045-008-0018-5

Tho YP (1987) Ecological and biological considerations in the management of the tropical rainforest ecosystem. In AMIC-CDG-COMCON-UKM Workshop on Mass Media and the Protection of the Environment: Kuala Lumpur, Asian Mass Communication Research and Information Centre, Singapore. https://hdl.handle.net/10356/101056

Timmer LW, Garnsey SM, Graham JH (2000) Compendium of citrus diseases. American Phytopathological Society, St Paul

Tripathi JN, Ntui VO, Malarvizhi M, Muiruri S, Ravishankar KV, Tripathi L (2023) Improvement of nutraceutical traits of Banana: new breeding techniques. In: Kole C (ed) Compendium of crop genome designing for nutraceuticals. Springer, Singapore

Tropical Fruits Network (2020) News compilation. https://www.itfnet.org/v1/news/tfnet-newsfeed/2020-2

Umetsu N, Shirai Y (2020) Development of novel pesticides in the 21st century. J Pestic Sci 45(2):54–74. https://doi.org/10.1584/jpestics.D20-201

Unnevehr L, Grace D (2013) Aflatoxins: finding solutions for improved food safety Focus 20. Brief 1 Nov 2013, IPFRI CGIAR, Research program on Agriculture for Nutrition and Health

Uppala S, Zhou XG (2018) Field efficacy of fungicides for management of sheath blight and narrow brown leaf spot of rice. Crop Prot 104:72–77

van den Hoogen J et al (2019) Soil nematode abundance and functional group composition at a global scale. Nature 572(7768):194–198. https://doi.org/10.1038/s41586-019-1418-6

Viswanathan R, Malathi P, Chandran K, Gopi R (2017) Screening of sugar cane germplasm for red rot resistance. J Sugarcane Res 7(2):100–111

Volynchikova E, Kim KD (2022) Biological control of oomycete Soilborne diseases caused by Phytophthora capsici, Phytophthora infestans, and Phytophthora nicotianae in Solanaceous crops. Mycobiology 50(5):269–293. https://doi.org/10.1080/12298093.2022.2136333

Wallace AR (1856). On the Bamboo and Durian of Borneo. Editor Charles H. Smith's Note: Included in a letter to Sir William Jackson Hooker; printed in Volume 8 of Hooker's Journal of Botany in 1856

Wang C, Zhuang W-Y (2018) Evaluating effective Trichoderma isolates for biocontrol of Rhizoctonia solani causing root rot of Vigna unguiculata. J Integr Agric 18(9):2072–2079

Wang KH, Uyeda J, Sugano MS (2018) Banana pest and disease management in the tropical pacific: a guidebook for banana growers. Chapter VI: IPM Strategies against Black Leaf Streak Disease. University of Hawaii at Manoa, p 72. https://cms.ctahr.hawaii.edu/wangkh/Research-and-Extension/Banana-IPM/Guidebook/CHPT6-IPM-BLS

Wang X, Yu R, Li J (2021) Using genetic engineering techniques to develop Banana cultivars with Fusarium wilt resistance and ideal plant architecture. Front Plant Sci 11:617528. https://doi.org/10.3389/fpls.2020.617528

Watanabe T, Igarashi H, Matsumoto K, Seki S, Mase S, Sekizawa Y (1977) Studies on rice blast controlling agent of benzisothiazole analogs. 1. Characteristics of probenazole (Oryzemate) for control of rice blast. J Pestic Sci 2:291–296

Weaver DB, Rodriguez-Kabana R (1992) Disease Management in Soybean: use of cultural techniques and genetic resistance. In: Copping LG, Green MB, Rees RT (eds) Pest Management in Soybean. Springer, Dordrecht. https://doi.org/10.1007/978-94-011-2870-4_21

Webb JLA (2002) Tropical pioneers: human agency and ecological change in the highlands of Sri Lanka. Ohio University Press, Athens

Wolfe MS (2000) Crop strength through diversity. Nature 406:681–682

Wolfe MS (2019) Agroforestry can decentralise production of food and energy and encourage 'commoning'. In: 4th world agroforestry congress, Montpellier

Wolfe MS, Barrett JA (1980) Can we lead the pathogen astray? Plant Dis 64:148–155

Woodward JE, Brenneman TB, Kemerait RC Jr (2013) Chemical control of peanut diseases: targeting leaves, stems, roots, and pods with foliar-applied fungicides. In: Fungicides–showcases of integrated plant disease management from around the world, vol 15, pp 55–76

Xie H, Lin C, Lu W, Han Z, Wei D, Huo X, Li T et al (2023) OsBLS6.2: a rice bacterial leaf streak resistance gene identified by GWAS and RNA-seq. Crop J 11(6):1862–1871. https://doi.org/10.1016/j.cj.2023.08.007

Yuan S, Linquist BA et al (2021) Sustainable intensification for a larger global rice bowl. Nat Commun 12:7163. https://doi.org/10.1038/s41467-021-27424-z

Zakaria L (2022) Fungal and oomycete diseases of minor tropical fruit crops. Horticulturae 8(4):323. https://doi.org/10.3390/horticulturae8040323

Zakaria L (2023) Basal stem rot of oil palm: the pathogen, disease incidence, and control methods. Plant Dis 107:603–615. https://doi.org/10.1094/PDIS-02-22-0358-FE

Zhang S, Moyne A-L, Reddy MS, Kloepper JW (2002) The role of salicylic acid in induced systemic resistance elicited by plant growth-promoting rhizobacteria against blue mold of tobacco. Biol Control 25(3):288–296. https://doi.org/10.1016/S1049-9644(02)00108-1

Zhao S, Jang S, Lee YK, Kim DG, Jin Z, Koh HJ (2020) Genetic basis of tiller dynamics of rice revealed by genome-wide association studies. Plants (Basel) 9(12):1695. https://doi.org/10.3390/plants9121695

Zhao C, Li Y, Liang Z, Gao L, Han C, Wu X (2022) Molecular mechanisms associated with the resistance of Rhizoctonia solani AG-4 isolates to the succinate dehydrogenase inhibitor Thifluzamide. Phytopathology 112:567–578

Zhi H, Wang D (2022) Chapter four – inheritance and prevention of soybean root rot. Adv Bot Res 102:105–119

Zhou XW, Su KQ, Zhang YM (2015) Phylogenetic analysis of widely cultivated *Ganoderma* in China based on the mitochondrial V4-V6 region of SSU rDNA. Genet Mol Res 14(1):886–897

Zhu Y, Chen H, Fan J, Wang Y, Li Y, Chen J, Fan J, Yang S, Hu L, Leung H, Mew TW, Teng PS et al (2000) Genetic diversity and disease control in rice. Nature 406(6797):718–722. https://doi.org/10.1038/35021046

Disclaimer

The information presented herein in this book represents the views of the author as of the date of publication. Because of the rate at which conditions change, the author reserves the right to alter and update his opinions based on new conditions. This document is for informational purposes only, and the author does not accept any responsibility for any liability resulting from the use of this information. While every attempt has been made to verify the information provided here, the author, resellers and affiliates cannot assume responsibility for errors, inaccuracies or omissions. Any slights of people or organizations are unintentional.

Index

A
Aflatoxins, 29, 38
Agroecosystems, 86–88
Agronomic measures, 28, 31, 96
Alternaria brassicicola, 71
Alternaria padwicki, 24
Alternaria porri, 69, 71
Alternaria solani, 68
Anthracnose disease, 75
Aphid, 77
Artificial intelligence (AI), 8, 91–92
Asian soybean rust (ASR) disease, 4, 39, 41

B
Bacterial canker disease, 68
Bacterial leaf blight disease, 26
Bacterial leaf streak diseases, 25
Basal stem rot disease, 56
Basidiocarps, 56, 58, 60
Biocontrol, 68, 88, 97, 102
Biodiversity, 4, 6, 8, 85–87, 95
Biological control agents, 97–102
Biostimulants, 88, 91, 92
Biotroph, 31
Bipolaris oryzae, 2, 25
Black death, 51
Black pod disease, 59–61
Black shank, 36, 37, 44
Black shank disease, 36, 37, 44
Black sigatoka disease, 51
Black swans, 9, 96
Blast disease, 21–23, 27, 97, 103
Blue mould disease, 36

Breeding, 26, 28, 50, 69, 76, 87, 90
Brown leaf hoppers, 26
Brown root diseases, 57
Brown spot disease, 2, 25
Burley tobacco, 35

C
Candidatus Liberibacter asiaticus, 4
Cassava mosaic virus disease, 42
Cavendish bananas, 47
Cercospora oryzae, 24
Cercospora spp., 42
Clavibacter michiganensis, 30, 68, 87
Climate changes, 1, 6, 8, 9, 40, 81–92
Cochliobolus heterotrophus, 31
Cochliobolus miyabeanus, 24, 25
Colletotrichum falcatum, 101, 105
Colletotrichum gloeosporiodes, 66, 67
Common smut disease, 30, 32
Complexity, 1, 6, 8–9, 75, 81, 83, 87, 88
Consultative Group for International Agricultural Research (CGIAR), 9, 27, 91
Cordana leaf spot disease, 50
Cordana musae, 52
Crazy top disease, 30, 32
Crinipellis perniciosa, 3
Crop diversification, 86
Crop domestication, 6, 7, 39
Crop mixtures, 85, 87, 89
Crop trends, 14–17
Crown blight disease, 66, 67
Crown rot disease, 54

Cryptic disease, 6
Cucurbit vegetables, 70–71
Cynefin framework, 83

D
Damping off disease, 37
Diagnostics, 42, 56–57, 88, 90
Didymella bryoniae, 70
Dieback disease, 59, 60, 74, 76
Dirty panicle disease, 24, 25, 27
Disease trends, 13–17
Disease triangle, 5, 83
DNA, 5, 55, 56
Downy mildew diseases, 30, 32

E
Early blight disease, 68
Ecosystem services, 85, 86, 96
Erwinia chrysanthemi, 77
Erysiphe cichoracearum, 69, 71
Erythricium (*Corticium*) *salmonicolor*, 59

F
False smut disease, 24, 27
Food security, 1, 6, 8, 9, 16, 29, 39–41, 81–92
Foot rot disease, 67, 70, 77
Forest rental, 86
Founder principle, 82
Freight train trends, 9, 81–82, 91–92
Frosty pod rot disease, 3, 59
Fruit rot disease, 66, 76, 78
Fungicide resistance, 88
Fungicide Resistance Action Committee (FRAC), 97
Fusarium oxysporum, 3, 49, 52, 53, 56, 67, 70
Fusarium oxysporum f.sp. *elaedis* (FOE), 56

G
Ganoderma boninense, 55, 56
Ganoderma lucidum, 55, 56
Ganoderma pseudoferreum, 57
Gene pyramiding, 22
Genetic origin, 5
Genetic resources, 90
Genomes, 31, 50, 90, 91
Genomics, 39, 42, 48, 66, 90, 91, 95

Grafting, 66, 70
Green leaf hoppers, 26, 28
Growth rings, 84
Gummy stem blight disease, 67, 70

H
Hemileia vastatrix, 3
Host resistance, 22, 23, 31, 41, 42, 50, 68, 69, 76, 87, 88, 90, 102
Huanglongbing disease, 4, 104
Hybrid rice, 16, 20, 27–28

I
Integrated Pest Management (IPM), 28, 86, 88–89, 96, 97

J
Java downy mildew disease, 30

K
Kernel smut diseases, 24, 25

L
Late blight disease, 2
Leaf blight disease, 57
Leaf fall of rubber, 57
Leaf spot diseases, 24
Legumes, 16, 35, 39, 41, 42, 54, 59
Lethal yellowing disease, 56
Lingzhi, 55, 56
Loop-mediated isothermal amplification (LAMP), 56–57, 90

M
Magnaporthe oryzae, 21, 22, 28, 82
Magnaporthe salvinia, 25
Meloidogyne spp., 68
Microbial plant defence elicitors, 101
Miracle rice, 16
Modelling, 28, 81–83, 90
Molecular methods, 95
Moniliophthora rorei, 3
Mosaic virus disease, 77
Mutants, 31
Mycosphaerella fijiensis, 50, 51
Mycosphaerella musicola, 50–52

Index 123

N
Narratives, 2, 8–10, 13, 95–96
Narrow brown spot disease, 24
Nematodes, 53, 60, 68, 69, 88

O
Oideum neolycopersici, 68
Oncobasidium theobromae, 60, 61

P
Panama disease, 3, 4, 16, 48, 49, 52–54, 83
Passalora arachidicola (formerly *Cercospora arachidicola*), 44
Patch canker disease, 79
PCR, 5, 26, 56, 90
Peronosclerospora maydis, 30
Peronosclerospora philippinensis, 30, 32
Peronosclerospora sacchari, 30
Peronosclerospora sorghi, 30
Peronospora hyoscyami f. sp. *Tabacina*, 36
Phakopsora pachyrhizi, 4, 41
Phellinus noxius, 57
Philippine downy mildew disease, 30
Phytophthora capsici, 66, 67
Phytophthora cinnamomi, 100
Phytophthora infestans, 2, 68
Phytophthora nicotianae, 37, 44
Phytophthora palmivora, 57, 59–61, 76, 79
Phytoplasma palmae, 4
Phytosanitation, 82
Pink disease, 59
Plant defence inducers, 101
Podosphaera, 98–101, 104
Post-harvest, 49, 53, 54, 74
Powdery mildew disease, 42, 68, 69, 71, 89
Predictive and decision models, 92
Problem based learning, 96
Proteomics, 31
Pseudocercospora fijiensis, 50, 51
Pseudocercospora musicola, 50–52
Pseudocercospora ulei, 3
Pseudoperonospora cubensis, 70
Purple blotch disease, 69, 71
Pythium, 30, 77

R
Red root disease, 57–58
Red rot disease, 101, 105
Regenerative agriculture, 84, 85, 88

Resilience, 2, 9, 40, 81–92, 95, 96
Rhizoctonia blight disease, 68
Rhizoctonia solani, 23, 24, 30, 42
Rice tungro virus (RTV), 26
Rigidoporus lignosus, 57, 58
Ring spot virus (RSV-P), 77
Ring spot virus disease, 77
Rootknot disease, 68, 69
Root rot disease, 30, 57–58, 76

S
Sanitation, 25, 28, 30, 37, 53, 56, 68, 76, 102
Sclerophthora macrospora, 30
Seed treatment, 24, 30
Sesame, 69, 71
Sheath blight disease, 23–26, 42
Solanaceous vegetables, 65–69, 71
Surveillance, 90
Survival, 1, 7, 23, 24, 26, 38–39, 41, 59, 95
Sustainability, 1, 6, 8, 9, 16, 40, 81–92, 95, 96
Sustainable agriculture, 16
Systemic acquired resistance (SAR), 5

T
Temperate, 6, 7, 19, 29, 30, 69, 84
Thanatephorus cucumeris (Rhizoctonia solani), 23, 24
Tilletia barclayana, 27
Transcriptomics, 90
Tungro virus disease, 28

U
Ustilago maydis, 30–32
Ustilginoidea virens, 27

V
Vascular streak dieback (VSD), 6, 59–61, 90
Vectors, 1, 4, 26, 42, 81, 83, 97
Vector control, 97
Virginia tobacco, 35
Viruses, 5, 16, 26, 28, 40, 77, 90

X
Xanthomonas campestris pv *oryzae*, 25, 26
Xanthomonas campestris pv *oryzicola*, 25
Xylella fastidiosa subsp. Pauca, 4

GPSR Compliance

The European Union's (EU) General Product Safety Regulation (GPSR) is a set of rules that requires consumer products to be safe and our obligations to ensure this.

If you have any concerns about our products, you can contact us on

ProductSafety@springernature.com

In case Publisher is established outside the EU, the EU authorized representative is:

Springer Nature Customer Service Center GmbH
Europaplatz 3
69115 Heidelberg, Germany

www.ingramcontent.com/pod-product-compliance
Lightning Source LLC
LaVergne TN
LVHW011000250326
834688LV00003B/40

9783031907890